SpringerBriefs in Applied Sciences and Technology

SpringerBriefs present concise summaries of cutting-edge research and practical applications across a wide spectrum of fields. Featuring compact volumes of 50 to 125 pages, the series covers a range of content from professional to academic.

Typical publications can be:

- A timely report of state-of-the art methods
- An introduction to or a manual for the application of mathematical or computer techniques
- A bridge between new research results, as published in journal articles
- A snapshot of a hot or emerging topic
- An in-depth case study
- A presentation of core concepts that students must understand in order to make independent contributions

SpringerBriefs are characterized by fast, global electronic dissemination, standard publishing contracts, standardized manuscript preparation and formatting guidelines, and expedited production schedules.

On the one hand, **SpringerBriefs in Applied Sciences and Technology** are devoted to the publication of fundamentals and applications within the different classical engineering disciplines as well as in interdisciplinary fields that recently emerged between these areas. On the other hand, as the boundary separating fundamental research and applied technology is more and more dissolving, this series is particularly open to trans-disciplinary topics between fundamental science and engineering.

Indexed by EI-Compendex, SCOPUS and Springerlink.

More information about this series at http://www.springer.com/series/8884

Danial Jahed Armaghani · Aydin Azizi

Applications of Artificial Intelligence in Tunnelling and Underground Space Technology

 Springer

Danial Jahed Armaghani
Department of Civil Engineering
Faculty of Engineering
University of Malaya
Kuala Lumpur, Malaysia

Aydin Azizi
School of Engineering
Computing and Mathematics
Oxford Brookes University
Wheatley Campus
Oxford, UK

ISSN 2191-530X   ISSN 2191-5318 (electronic)
SpringerBriefs in Applied Sciences and Technology
ISBN 978-981-16-1033-2   ISBN 978-981-16-1034-9 (eBook)
https://doi.org/10.1007/978-981-16-1034-9

This Springer imprint is published by the registered company Springer Nature Singapore Pte Ltd.
The registered company address is: 152 Beach Road, #21-01/04 Gateway East, Singapore 189721, Singapore

About This Book

This book covers the tunnel boring machine (TBM) performance classifications, empirical models, statistical, and intelligent-based techniques, which have been applied and introduced by the previous researchers in this field. In addition, a critical review of the available TBM performance predictive models will be discussed in detail. Then, this book introduces several predictive models, i.e., statistical and intelligent techniques, which are applicable, powerful, and easy to apply, in estimating TBM performance parameters. The introduced models are accurate enough and they can be used for the prediction of TBM performance in practice before designing TBMs.

Contents

1 An Overview of Field Classifications to Evaluate Tunnel Boring
Machine Performance 1
  1.1 Introduction 1
  1.2 Tunnel Boring Machine 2
     1.2.1 Brief History of TBM 3
     1.2.2 Types and Basic Principles of TBM 3
     1.2.3 TBM Performance Parameters 5
     1.2.4 Factors Influencing TBM Performance 7
  1.3 TBM Prediction Field Classifications 8
  1.4 TBM Performance Prediction Using Field Approach 10
  1.5 RMCs Used in TBM Performance Prediction 11
  1.6 Discussion and Conclusion 13
  References 13

2 Empirical, Statistical, and Intelligent Techniques for TBM
Performance Prediction 17
  2.1 Introduction 17
  2.2 Theoretical Models 18
     2.2.1 Cutter Load Approach 19
     2.2.2 Specific Energy Approach 20
  2.3 Empirical Models 21
  2.4 Statistical Approach 22
  2.5 Computational-Based Techniques 24
  2.6 Discussion and Conclusion 25
  References 28

3 Developing Statistical Models for Solving Tunnel Boring Machine
Performance Problem 33
  3.1 Introduction 33
  3.2 Regression-Based Models 34
     3.2.1 Linear Multiple Regression (LMR) 34
     3.2.2 Non-linear Multiple Regression (NLMR) 35
  3.3 Case Study 35

3.4   Data Measurement and Input Variables 36
      3.4.1   Rock Material Properties 37
      3.4.2   Rock Mass Properties 37
      3.4.3   Machine Characteristics 38
      3.4.4   Input Variables 38
3.5   Regression-Based Models 39
      3.5.1   Simple Regression 39
      3.5.2   Multiple Regression 40
3.6   Discussion and Conclusion 44
References 49

4   A Comparative Study of Artificial Intelligence Techniques
    to Estimate TBM Performance in Various Weathering Zones 55
    4.1   Introduction 55
    4.2   Methodology 56
          4.2.1   Artificial Neural Network (ANN) 56
          4.2.2   Group Method of Data Handling (GMDH) 57
    4.3   Tunnel Site and Data Collection 58
    4.4   GMDH Model Development 59
    4.5   Model Assessment and Discussion 63
    4.6   Conclusions 66
    References 66

About the Authors

Dr. Danial Jahed Armaghani is currently working as a senior lecturer in the Faculty of Engineering, University of Malaya, Malaysia. He received his postdoc from Amirkabir University of Technology, Tehran, Iran and his Ph.D. degree, in Civil-Geotechnics, from Universiti Teknologi Malaysia, Malaysia. His area of research is tunnelling, rock mechanics, piling technology, blasting environmental issues, applying artificial intelligence, and optimization algorithms in civil-geotechnics. Dr. Danial published more than 150 papers in well-established ISI and Scopus journals, national, and international conferences. Dr. Danial is also a recognized reviewer in the area of rock mechanics and geotechnical engineering.

Dr. Aydin Azizi holds a Ph.D. degree in Mechanical Engineering. Certified as an official instructor for the Siemens Mechatronic Certification Program (SMSCP), he currently serves as a Senior Lecturer at the Oxford Brookes University. His current research focuses on investigating and developing novel techniques to model, control, and optimize complex systems. Dr. Azizi's areas of expertise include Control and Automation, Artificial Intelligence and Simulation Techniques. Dr. Azizi is the recipient of the National Research Award of Oman for his AI-focused research, DELL EMC's "Envision the Future" completion award in IoT for "Automated Irrigation System", and "Exceptional Talent" recognition by the British Royal Academy of Engineering.

Chapter 1
An Overview of Field Classifications to Evaluate Tunnel Boring Machine Performance

Abstract As a difficult and complex task, the accurate prediction of the tunnel boring machine (TBM) performance in various geological/ground conditions is of great importance and interest. Over the last decades, many rock mass classifications and field approaches have been developed to predict TBM performance in a reliable way. This study gives an overview of the mentioned models and their performance capacity in estimating TBM performance in different conditions. The review of rock mass classifications and field approaches indicated that these are considered as site-specific techniques and the performance prediction of these techniques is not satisfactory. In addition, these techniques are complex with many predictors or input parameters while providing all input parameters is sometimes impossible or very difficult for a specific tunnelling project. This research suggests other techniques such as statistical-based and computational-based in order to get a higher level of accuracy in the area of TBM performance.

Keywords Tunnel boring machine · TBM performance · Rock mass classification · Field approach

1.1 Introduction

Developed in recent decades, tunnel boring machine (TBM) is an equipment designed for the construction of tunnelling projects. TBM is now a standard method of excavating tunnels with lengths over 1.5–2 km [1]. Several factors including economic considerations and schedule deadlines govern its application in civil and mining projects [2–4]. TBMs have been used to excavate tunnels with various mass conditions such as blocky and weathered grounds [5, 6].

James S. Robbins has improved TBM design several times since its first introduction in 1954 so that it could be utilized to a greater range of rock mass and material conditions with a better performance. Thus, more powerful and more efficient TBMs have been developed to be utilized with a wide variety of hard and soft rocks. In this regard, a challenging issue that comes up is the prediction of TBM's performance in difficult rock masses. Valuable information can be obtained about

© The Author(s), under exclusive license to Springer Nature Singapore Pte Ltd. 2021
D. Jahed Armaghani and A. Azizi, *Applications of Artificial Intelligence in Tunnelling and Underground Space Technology*, SpringerBriefs in Applied Sciences and Technology, https://doi.org/10.1007/978-981-16-1034-9_1

the geological conditions before tunnelling and the rock mass response to excavation through geological documentation [7]. TBM performance is commonly affected by variables such as geological conditions and weathering, mass and material strength, and machine factors [8–10]. Planning tunnel projects is a tough task that necessitates predicting TBM performance in order to select proper methods of construction. Good plans or techniques can highly reduce the risks connected with capital costs; a common point of concern in the tunnel excavation business [11, 12].

Many researchers [5, 13–18] found that TBM performance depends on three major groups; (1) machine factors, (2) mass properties of the rock, and (3) material properties of the rock. In many of the early conducted research in this area, the authors consider one of two important factors from the mentioned groups. So far, many models and classifications have been developed to forecast TBM performance. Barton [5] developed the Q_{TBM} model based on the Q-system [19] in order to forecast the penetration rate (PR) and advance rate (AR) of TBM. Compared to Q-system, Q_{TBM} takes more parameters into consideration in TBM applications. As another model, rock mass excavatability (RME) was introduced by Bieniawski et al. [20] for the prediction of TBM AR. Till now, the RME index has gone through several revisions. Several researchers have used the Q_{TBM} and RME models for the estimation of TBM PR and AR in their case studies [21].

In this research, an overview of rock mass classifications (RMCs) and field approaches will be given to estimate TBM performance in various ground conditions. The advantage and disadvantage of the mentioned models/approaches will be discussed, and the best models among them will be selected. In addition, the most important parameters on TBM performance will be described in this study.

1.2 Tunnel Boring Machine

TBM is an excavation machine cutting the rock full-face by pushing and rotating the cutter head. According to Goel and Singh [22], TBM has extreme rates of tunnelling of 15 km/year and 15 m/year and occasionally even less. Basically, excavation by TBMs has three advantages, namely, negligible disturbance for the adjoining rock mass, high safety with low overbreaks, and low manpower [23]. On the other hand, it takes a significant time for setup and dismantling, and there is a limited range of shapes for the available tunnel cross-section [24]. To expect fast progress of tunnel construction, it is important to have the evaluation of hydrogeology and geology conditions alongside the route of a tunnel. Since TBMs are capable of cutting rock up to 300 MPa, they can be applied to most hard rock [23].

TBM can be generally described as a tool equipped with disc cutters used to drive tunnels. Cutting rocks are made possible by this excavation machine when the cutter head rotates and the blade exerts pressure on the face. Another definition offered for a TBM has been that of a milling machine; a definition that does not give an account of how the machine operates. The application of TBMs is different from drilling and blasting operations in that it does not provide the possibility of a flexible reaction to the tunnel–rock interaction through subdividing the excavated section or rapidly adapting the support to the geological conditions.

1.2.1 Brief History of TBM

Early tunnelling machines could not be considered as TBMs since these conventional machines did not operate on the whole face using their excavation tools. Rather, they broke out a groove around the tunnel wall. Having cut the groove, the machine was withdrawn and explosives or wedges were used to loosen the remaining core. In 1851, an American engineer called Charles Wilson developed and built a TBM and patented it in 1856. Having all the properties that a modern TBM possesses, Wilson's machine can be classified as the first machine working by boring the tunnel. Before building it, Wilson developed disc cutters (for which he also applied for a patent) in 1847 so that the machine would be able to excavate the entire face [23].

An entirely new system was proposed by Cooke and Hunter (Wales) with their patent in 1866 [25]. A large drum was placed in the centre and moved ahead of the outer drums, which were smaller in diameter and extended the cross-section. The excavated section was in the shape of a box and extended at right angles. In 1863, Frederick Beaumont applied for a patent for a tunnelling machine equipped with chisels but the machine failed to construct a water tunnel. In 1875, he built a tunnel boring machine, which had a rotating cutting wheel and applied for another patent [25]. In 1931, gallery cutting machines built by Schmidt, Kranz and Co. came along as a new and more successful generation. In this generation, the equipment includes cable carriage, drill carriage, loading band, and bracing carriage as the main apparatuses [26].

In the 1950s, a breakthrough occurred in the development of modern TBMs. James S. Robbins, who was a mine engineer, was successfully developed the first open gripper TBM consisted of disc cutters as its sole tool. Early experiments with a machine that had only disc cutters, conducted in driving the Humber sewer tunnel in Toronto, revealed that this type could achieve the same advance performance as the former types of TBM that combined hard metal cutters and discs. In addition, the new machine's working life was significantly longer.

1.2.2 Types and Basic Principles of TBM

In rock mass ground and especially in hard rock condition, there are various types of tunnelling machines as shown in Fig. 1.1 [27]. Based on this figure, there are two types of full-face TBMs, namely, gripper TBMs and shield TBMs. In the following

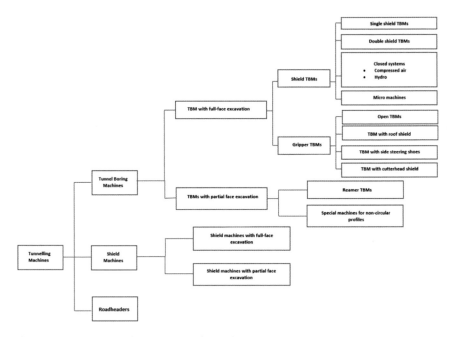

Fig. 1.1 Different types of tunnelling machines [27]

paragraphs, basic principles of gripper TBMs (as the most important TBM type) are described. The descriptions regarding other types of TBMs can be found in the literature [23].

The gripper TBM, also named open TBM, is known as a typical form of the tunnelling machine. The gripper TBM can be used in hard rock condition with stand-up time ranging from medium to high. The use of this form can be the most economical in cases where the rock does not require a constant support with the steel arches, rock anchors or even shotcrete.

Sub-categories of gripper TBMs:

Open TBM: This type of TBM does not have static protection units behind the cutter head. Today, only the small diameter version of this machine is available.

TBM with roof shield: It is very similar to the open type. In this type, if we have isolated rockfalls during excavation, the machine uses a system, namely, static protection roofs, which are located behind the cutter head to for protecting the tunnel crew.

TBM with roof shield and side steering shoes: In addition to the protection function, this type has side steering shoes in order to support the front when the machine is moving and steering during the boring operation. Radial driving against the tunnel walls is made possible through the side surfaces.

TBM with cutter head shield: The function of the cutter head in this type of machine is to protect the crew in the cutter head area. The short shield liner provides forward support as the machine moves.

The basic structure of a TBM consists of several important systems including boring, thrust and clamping, muck removal, and support, which will be discussed in the following [28].

Boring system, which is considered as the most influential component of TBM, includes a house for TBM disc cutters. In this house, the arrangement of discs is done in such a way that they can contact the whole cutting face. The difficulty of cutting and the type of rock determine how the cutting tracks are divided, and what discs are selected, which in turn determine what size the broken pieces of rock would have.

The thrust and clamping system: This system is an influential element in a TBM's performance. It is responsible for the progress of the boring process and brings about advance thrust. Hydraulic cylinders to thrust the cutter head with its drive unit forward apply the required pressure. The maximum stroke depends on the length of the thrust cylinder piston.

Muck removal system: A powerful system, which has no interference with the supply of the TBM and essential support measures, must be selected to ensure that the muck is carried away through the whole tunnel. Either a conveyor or a rail system can be used in accordance with the local conditions. Large dump trucks can also be used.

Support system: The only place for the support measures is the rear carriage behind the TBM. A problem always arises in the case of boring through fault zones with poor geology, where the advance time is longer than the rock's stand-up time [15, 29]. To enable the installation of ground support, drilling guides can be applied behind the cutter head (e.g., rock bolts). Expanding rings and shotcrete can also be erected behind the cutter head.

1.2.3 TBM Performance Parameters

Several parameters must be taken into account for evaluating the performance of a TBM system. Typically, PR, AR, and utilization index (UI) are measured as TBM performance parameters [30]. In the following, some of the most important are.

Utilization index (UI), expressed in percent, denotes the percentage of time in boring (T_b) per unit shift time (T_{sh}) [31]. Actually, T_{sh} is defined as a combination of T_b and the time wasted by other delays (T_d) throughout the tunnelling excavation:

$$UI = \frac{TBM\ boring\ time}{Shift\ time} \times 100 = \frac{T_b}{T_{sh}} \times 100 \qquad (1.1)$$

$$T_{sh} = T_b + T_d \qquad (1.2)$$

where, T_b is the cutter head operation time, expressed in hour.

Penetration rate (PR) signifies the ratio of excavation distance to the operating time for the duration of the tunnel construction process:

$$PR = \frac{Distance\ bored}{TBM\ boring\ time} = \frac{L}{T_b} \quad (1.3)$$

where PR, which is expressed normally in m/h or mm/min, stands for the average penetration rate. PR can also be measured according to the distance bored per cutter head revolution, which is defined as an instantaneous penetration or as an average over each thrust cylinder cycle. Penetration per cutter head revolution, P_{rev}, can be used to investigate the mechanics of rock cutting:

$$P_{rev} = \frac{1000 \times PR}{60 \times RPM} \quad (1.4)$$

where, P_{rev} is penetration per revolution of cutter head expressed in mm/rev, and RPM is the rate of cutter head revolutions per minute expressed in rev/min.

The relationship between specific energy (SE) and P_{rev} is suggested by Snowdon et al. [32] and Sanio [17] as follows:

$$SE = \frac{200 \times n_c \times r_c \times F_r}{3 \times D \times P_{rev}} \quad (1.5)$$

In the above equation, n_c denotes the number of cutters on the cutter head, r_c shows the weighted average cutter distance from the rotation centre in m, F_r signifies the cutter rolling force in kN, and D expresses TBM diameter in m. SE manifests the energy required by the disc cutters for cutting the rock unit volume, expressed in MJ/m^3.

The average speed of tunnel advancement is defined as advance rate (AR), which is normally denoted as m/day or m/h. AR represents the distance bored according to shift time.

$$AR = \frac{distance\ bored}{shift\ time} = \frac{L}{T_{sh}} \quad (1.6)$$

If PR and UI are expressed on the same time basis, AR is equated to:

$$AR = \frac{PR \times UI}{100} \quad (1.7)$$

AR varies based on the change in the parameters that affect either the PR or the UI. Those parameters include facing a very poor or hard rock, unstable invert that can bring about train derailments, creating an appropriate resistance against the grippers, reduction in torque capacity, TBM maintenance, highly abrasive rock causing the cutter to wear faster, break down, and tunnel collapse.

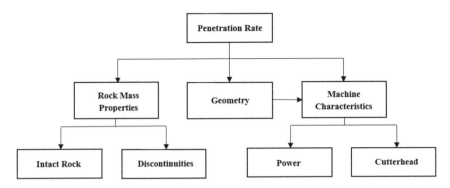

Fig. 1.2 Main parameters influenced the PR [18]

1.2.4   Factors Influencing TBM Performance

During the operations of rock excavation machines, several factors are involved, including intact rock properties, rock mass characteristics, machine characteristics, geological setting, expert knowledge, and operator skills [3, 6, 33–38].

Figure 1.2 displays the most important factors influencing TBM PR, which are tunnel geometry, machine factors, and rock material and mass properties [39].

Rock mass and material properties: The intact rock and the discontinuity structure of rock mass are used for the determination and evaluation of rock mass properties. According to the previous studies, rock compressive strength, known as UCS is the main parameter that influences PR results for intact rock [12, 40, 41]. Greater rock strength generally results in a lower PR. Bruland [42] stated that the discontinuities orientation can have an effect on PR. It is noticeable that other factors, e.g., discontinuity orientation, may also be important to the overall performance of TBM; thus these factors should be taken into account [18].

Machine characteristics: Grima et al. [18] state that the most important feature of a TBM is its thrust force. The number of cutters mounted on TBM can compute maximal thrust per cutter and the majority of the models has employed this as a measure. Raising the thrust per cutter is highly advantageous for the PR. Moreover, the maximal torque, maximal power, and maximal RPM (revolutions per minute) are related to thrust force. Primarily, these parameters are dependent upon the diameter of tunnel. The maximal levels are obtained at some definite sections throughout the construction process of tunnel, which introduces a large source of variance.

Geometry: Tunnel geometry is considered as a very important parameter. According to Grima [18], tunnel diameter is considered as an effective tunnel geometry parameter and some parameters such as RPM and torque can be affected by the mentioned factor.

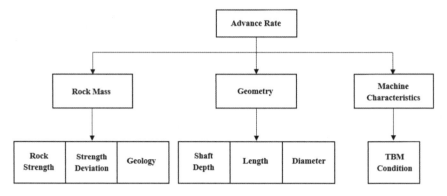

Fig. 1.3 Main parameters influenced the AR [18]

TBM excavation includes some activities, namely, repair and maintenance, routine boring, travel time, ground support placement, etc. Any changes in machine assumptions and anticipated geological conditions may influence the performance of TBM [31]. Armaghani et al. [30] and Grima et al. [18] stated that the AR of TBM can be influenced by three main categories, which are rock mass and material parameters (e.g., rock strength, strength deviation, and geology), machine characteristics (i.e., TBM condition) and geometry (e.g., shaft depth, length, and diameter). Figure 1.3 shows the effective parameters of AR.

1.3 TBM Prediction Field Classifications

Field studies have the merit of accounting for the complexity of factors related to the machine and geology, as well as those having to do with rock mass properties. Thus, the field approach is of interest to tunnel design engineers and project planners due to its being practical and based on the experiments conducted in actual tunnelling operations. In addition, the data gathered through such studies can be used to confirm and validate related experiments with disc cutters performed in the laboratory. Some models provided by field studies are discussed in the following.

NTNU model: Bruland [42] updated the model proposed by Lislerud [43], developed by the same Norwegian research group. Discontinuity spacing and direction, machine characteristics such as thrust per cutter, RPM, and cutter size are taken into consideration by this model. The model was developed through multivariable regression, which employs charts for defining the working parameters. The brittleness test and the Siever's miniature drill test were conducted to obtain the drilling rate index (DRI). Test procedures were elaborated by Bruland [44], including the DRI values of more than 2000 sample locations (approximately 80% of the data were collected in Norway). Spacing of the weakness planes has a considerable effect on the PRs, and difference of scale that exists between the actual cutters and point load tester became important.

RMi model: Another model that adopted the rock mass index (RMi) as its basis was proposed by Palmstrom [45], having the closest relation to the parameters of the NTNU model. In this model, rock mass factors, particularly the jointing properties have been properly taken into account. RMi specification of jointing and joints is composed of three-dimensional occurrences. Thus, it includes the influence of more than one joint set. In addition, the RMi parameters comprise joint features of significance for the shear strength of the joints that is effective on the TBM boring rate. For that reason, RMi must suitably assess the boring penetration of the tunnel in moderately hard and hard rock masses. In general, it is said that the presence of joints makes an improvement on the boring rate. Yet, due to a conservatism approach adopted in most analyses, boring rate improvement caused by jointing has been neglected in testing intact types of solid rock since predictions have been made on the basis of the strength characteristics of the intact rock.

Q_{TBM} model: Barton [5] developed Q_{TBM} as a new model based on the Q-system, which has already classified rock mass [19] but the former includes more machine–rock mass interaction parameters. Some of the delays involved can be explained by the conventional Q-value as well as the cutter life index (CLI) and quartz content. Q_{TBM} was proposed in order to predict PR and AR of the TBM. In addition to the existing parameters in Q-system including rock quality designation (RQD), number of joint set, roughness condition, alteration condition, water condition, and stress reduction factor, Barton [5] utilized other parameters, namely, average cutter load (F), compressive and tensile rock mass strength (σ_{cm} or σ_{tm}), CLI, the quartz content, biaxial stress on the tunnel face (σ_{θ}) in his proposed model. More details regarding Q_{TBM} model can be found in its original published research [5].

RME model: Bieniawski et al. [20] developed rock mass excavatability (RME) model, which is similar to the RMR classification system to predict the performance of double-shield and open-type TBMs. Excavatability refers to the excavation rate that is stated as machine performance in meters per day. The index has been reconsidered and modified by several researchers [46, 47]. Barton and Bieniawski [48] indicated that the development of RME was on the basis of the RMR classification, like Q_{TBM} that was developed using Q-system. When RME index is available, the tunnelling technique is selected based on the TBM performance quantification that includes the average rate of advance (ARA). Therefore, the input parameters of RME are those that have higher effect on ARA, including drillability, stand-up time, UCS of intact rock, groundwater inflow, and discontinuities at the excavation front. It is worth mentioning that using RME model, only results of actual AR can be predicted empirically. More details about RME model and its calculation procedure are available in its original study by Bieniawski et al. [20].

1.4 TBM Performance Prediction Using Field Approach

Goel [49] used Q_{TBM} and RME models to predict TBM performance of the Himalayan tunnel, India. It was found that the actual PR and AR results are less than the estimated PR and AR values obtained by Q_{TBM} and RME models. Palmstrom and Broch [21], first, described Q-system history and its related parameters and then utilized Q_{TBM} model for the prediction of TBM performance. In their conclusion, they stated that Q_{TBM} is a complex model and is not suggested to be used in its current form.

In northern Italy, TBM performance of three tunnels (14 km in total) was assessed using Q_{TBM}, RMR and Q-system classifications in the study were conducted by Sapigni et al. [7]. They reported a reasonable correlation between the net PR and the RMR values; however, they came to the conclusion that the results were greatly spread, thus the results could not be completely used in a practical way for predicting the TBM performance. In terms of Q-system and Q_{TBM}, they reported poorer correlations with net PR in comparison with the RMR classification. In addition, they indicated that Q_{TBM} is slightly sensitive towards the PR, and correlation coefficient with the recorded data is worse than the traditional Q or other essential parameters such as the intact rock strength. Sapigni et al. [7] stated that the Q_{TBM} model reliability cannot be tested using only a single case; however, the mismatch underlines those problems that are concerned with prediction of the performance at times that several factors (e.g., human experience, characteristics of the system of machine and muck removal, and rock mass condition) are involved.

Ribacchi and Lembo-Fazio [41] used data of Varzo tunnel in the form of Q_{TBM} model. A relationship was applied between PR and Q_{TBM} values and coefficient of determination (R^2) value of their relationship was less than R^2 value of Barton's Q relationship. In general, they concluded that RMR classification can perform better in predicting TBM performance compared to Q_{TBM} model. A comparison between three TBM performance prediction models, namely, NTNU, Colorado School of Mines, CSM [13] and Q_{TBM} was made in the study by Hassanpour et al. [50] using 8.7 km data of Karaj water conveyance tunnel. Their results indicated that these models need some adjustments or correction factors in order to obtain more accurate TBM performance estimation. In addition, considering the tunnel data and using SPSS software, a multiple regression analysis was proposed to predict field penetration index (FPI). A R^2 value of 0.58 for the proposed equation shows a suitable accuracy of their proposed equation for the prediction of FPI. Furthermore, Hassanpour et al. [1] obtained similar results of the basic RMR, geological strength index (GSI), and Q in different points of Karaj water conveyance tunnel. They concluded that NTNU model indicates more reasonable results compared to CSM and Q_{TBM} models. In addition, the obtained values of PR using NTNU and Q_{TBM} models were closer to actual PR values in comparison to CSM model.

Oraee and Salehi [51] employed three experimental methods namely NTNU, Q_{TBM} and CSM to predict TBM PR using 16 km data of Karaj–Tehran water supply tunnel, Iran. They stated that Q_{TBM} is more suitable in tunnel projects with many joint structures in which expensive testing conducted is not possible. After a series of analyses, they concluded that Q_{TBM} model obtained higher credibility level than the other mentioned models in predicting TBM PR. Armaghani [52] used Q_{TBM} and RME as two field approaches to predict TBM performance of three TBMs in hard rock and weathered rock mass. He showed that these two techniques are not good enough for getting a high performance for PR and AR prediction. Due to that, he developed statistical and intelligent techniques for the estimation of TBM performance with a high level of accuracy.

1.5 RMCs Used in TBM Performance Prediction

Several systems for the classification of rock mass have been proposed in the fields of civil and mining engineering. The developed models have been often used in many empirical design practices in the area of tunnelling and rock mechanics. The RMC systems have been widely used for designing and construction of underground space, tunnelling, and rock-based structures. Since Ritter [53] made an attempt to formalize an empirical model for design of tunnel support, the schemes of RMC systems have been proposed for over 100 years.

Deere et al. [54] proposed the index of RQD for characterization and design. It was extended to present a quantitative estimation of the quality of rock mass from the drill logs. Later, other researchers introduced some other classifications of rock mass incorporating multiple parameters. Among the already-proposed classification systems, some have been commonly applied in mining and civil engineering fields, including the RMR by Bieniawski [55, 56], rock structure rating (RSR) by Wickham et al. [57], Q-system by Barton et al. [19] and geological strength index (GSI) by Hoek and Brown [58]. Many attempts have been made to show the relationships between different RMCs and TBM performance. Tallon [59] applied several classifications including RSR, Q-system, and RMR for predicting TBM performance of several tunnel projects in Northern Spain. The following conclusions can be drawn from his study:

(i)   the RSR system has a tendency to concentrate on the medium strength rocks but ignores the poor or strong ones,
(ii)   the Q-system provides the best definition over the whole range of rocks, in particular, on very weak rocks and
(iii) the RMR system makes a good distribution between medium and strong rocks but fails to make a good differentiation with the poor strength rocks.

Innaurato et al. [40] made an update on the method that had been already proposed by Cassinelli et al. [60]. Their method comprises RSR classification introduced by Wickham et al. [57]. The most important change in the updated version is that UCS was incorporated in the RMC. This is noticeable that the RSR was developed originally to determine the proper steel rib tunnel wall support, which included some parameters like rock type, joint spacing, joint condition, water inflow, geological structure, and dip direction. Within RSR method, the intact rock strength is partly taken into account by the rock type and classification by hardness. Probably, this is why there is UCS in the Innaurato's model. The method is based upon 112 homogeneous tunnel sections: however, no information is provided on the number of bored tunnels.

$$PR = \sigma_c^{-0.437} - 0.047\,RSR + 3.15 \qquad (1.8)$$

where RSR is rock structure rating, σ_c is uniaxial compressive strength of the intact rock (MPa).

Jalil et al. [61] employed RMR and Q-system parameters for TBMs performance prediction and stated that the Q value is more sensitive to variation in the extent of weathering, as reflected by changes in the alteration and roughness parameters. The RMR system is less sensitive to variation in joint surface condition. Sapigni et al. [7] carried out a study surveying a total of 14 km of tunnel almost continually, yielding more than 700 sets of data that featured the rock mass characteristics and the TBM performance. Because there is an empirical relation between PR and RMR, it can be said that TBM performance reaches a maximum in RMR range of 40–70; whereas a slower penetration was observed in both too good and too bad rock masses.

Ribacchi and Lembo-Fazio [41] made a correlation between the parameters of TBM performance, namely, PR, AR, and specific penetration to two rock quality indices, i.e., RMRP and joint spacing in a gneissic formation. They came to a conclusion that even the straightforward quality indices, e.g., partial rating of joint spacing in the RMR classification, have an acceptable predictive value for the PR. A list of RMC relationships to predict TBM performance parameters are presented in Table 1.1. Among these relationships, except the one presented by Ribacchi and Lembo-Fazio [41], most of the equations make direct use of the RMC systems as the main variable. Based on Table 1.1, minimum R^2 (0.369) and maximum R^2 (0.532) are obtained for Q-system and basic RMR, respectively. As discussed in the previous sections and according to the presented results in this table, RMCs can predict TBM performance with relatively low reliability.

Table 1.1 A list of RMCs relationships to predict TBM performance parameters

Reference	Correlation	R^2
Cassinelli et al. [60]	$PR = -0.0059\,RSR + 1.59$	–
Innaurato et al. [40]	$PR = \sigma_c^{-0.437} - 0.047\,RSR + 3.15$	–
Ribacchi and Lembo-Fazio [41]	$SP = 250\,\sigma_{cm}^{-0.66}$, $\sigma_{cm} = \sigma_c\,\exp\left(\frac{RMR-100}{18}\right)$	0.470
Hassanpour et al. [50]	$FPI = 0.222\,BRMR + 2.755$	0.532
	$FPI = 0.178\,RMR + 4.005$	0.449
	$FPI = 9.273\,e^{0.008GSI}$	0.412
	$FPI = 11.718\,Q^{0.098}$	0.369

PR = Penetration rate; σ_c = Compressive strength of rock material; SP = Specific penetration; σ_{cm} = Compressive strength of rock mass; FPI = Field penetration index; BRMR = Basic RMR

1.6 Discussion and Conclusion

Many field approaches and RMCs were proposed to solve problem related to TBM performance and its prediction before tunnel construction in various ground conditions. In case of field approaches, different researchers reported a wide range of model accuracy and suitability. Some of them mentioned that field approaches are complex and need to provide many predictors or parameters for predicting TBM performance. Some others reported a poor correlation of these approaches with TBM performance parameters. On the other hand, there are a few opinions highlighting a reliable and suitable performance prediction of these approaches. In general, it can be concluded that these approaches are not of interest and importance by researchers in this field since they need many parameters to be provided. Another shortcoming of these approaches is related to their low level of accuracy, which is a major issue in tunnelling projects.

In terms of RMCs and their use in tunnelling performance prediction, there are many published studies/formulas using and applying different RMCs such as RMR, RSR, and GSI. The accuracy of these models was examined using R^2 values and a range of 0.37–0.53 was achieved for R^2 of the proposed equations using different RMCs. The best R^2 was obtained using BRMR or basic RMR classification and the worst R^2 was achieved using Q-system classification. In general, it should be mentioned that these classifications are not able to provide a suitable and applicable accuracy level in predicting TBM performance.

Reviewing both field approaches and RMCs models showed that these techniques are site specific and required to provide many influential factors/parameters. On the other hand, their accuracy level in predicting TBM performance parameters is not satisfactory. Therefore, some other techniques that are statistical-based or computational-based may be useful to be utilized in the area of TBM performance [62–65]. Using the mentioned techniques, the performance prediction of TBM can be increased significantly compared to the reviewed approaches in this study.

References

1. J. Hassanpour, J. Rostami, J. Zhao, A new hard rock TBM performance prediction model for project planning. Tunn. Undergr. Sp. Technol. **26**, 595–603 (2011)
2. G. Girmscheid, C. Schexnayder, Tunnel boring machines. Pract. Period. Struct. Des. Constr. **8**, 150–163 (2003)
3. J. Zhou, Y. Qiu, S. Zhu, D.J. Armaghani, M. Khandelwal, E.T. Mohamad, Estimation of the TBM advance rate under hard rock conditions using XGBoost and Bayesian optimization. Undergr. Sp. (2020). https://doi.org/10.1016/j.undsp.2020.05.008
4. J. Zeng, B. Roy, D. Kumar, A.S. Mohammed, D.J. Armaghani, J. Zhou, E.T. Mohamad, Proposing several hybrid PSO-extreme learning machine techniques to predict TBM performance. Eng. Comput. (n.d.) (2021). https://doi.org/10.1007/s00366-020-01225-2
5. N. Barton, TBM perfomance estimation in rock using QTBM. Tunn. Tunn. Int. **31**, 30–34 (1999)

6. D.J. Armaghani, E.T. Mohamad, M.S. Narayanasamy, N. Narita, S. Yagiz, Development of hybrid intelligent models for predicting TBM penetration rate in hard rock condition. Tunn. Undergr. Sp. Technol. **63**, 29–43 (2017). https://doi.org/10.1016/j.tust.2016.12.009

7. M. Sapigni, M. Berti, E. Bethaz, A. Busillo, G. Cardone, TBM performance estimation using rock mass classifications. Int. J. Rock Mech. Min. Sci. **39**, 771–788 (2002)

8. S. Yagiz, Assessment of brittleness using rock strength and density with punch penetration test. Tunn. Undergr. Sp. Technol. **24**, 66–74 (2009)

9. A.C. Yagiz, S. Ghasemi, E. Adoko, Prediction of rock brittleness using genetic algorithm and particle swarm optimization techniques. Geotech. Geol. Eng. **36**, 3767–3777 (2018)

10. E. Ghasemi, H. Kalhori, R. Bagherpour, S. Yagiz, Model tree approach for predicting uniaxial compressive strength and Young's modulus of carbonate rocks. Bull. Eng. Geol. Environ. **77**, 331–343 (2018)

11. A.G. Benardos, D.C. Kaliampakos, Modelling TBM performance with artificial neural networks. Tunn. Undergr. Sp. Technol. **19**, 597–605 (2004)

12. M. Koopialipoor, H. Tootoonchi, D. Jahed Armaghani, E. Tonnizam Mohamad, A. Hedayat, Application of deep neural networks in predicting the penetration rate of tunnel boring machines. Bull. Eng. Geol. Environ. (2019). https://doi.org/10.1007/s10064-019-01538-7

13. J. Rostami, L. Ozdemir, A new model for performance prediction of hard rock TBMs, in *Proceedings Rapid Excavation and Tunneling Conference Society For Mining, Metallogy & Exploration* (Inc, 1993), p. 793

14. L. Ozdemir, F.-D. Wang, Mechanical tunnel boring prediction and machine design. Nasa Sti/Recon Tech. Rep. **80** (1979)

15. S. Yagiz, Development of rock fracture and brittleness indices to quantify the effects of rock mass features and toughness in the CSM Model basic penetration for hard rock tunneling machines (2002)

16. D.J. Armaghani, R.S. Faradonbeh, E. Momeni, A. Fahimifar, M.M. Tahir, Performance prediction of tunnel boring machine through developing a gene expression programming equation. Eng. Comput. **34**, 129–141 (2018)

17. H.P. Sanio, Prediction of the performance of disc cutters in anisotropic rock. Int. J. Rock Mech. Min. Sci. Geomech. Abstr. 153–161 (1985). (Elsevier)

18. M.A. Grima, P.A. Bruines, P.N.W. Verhoef, Modeling tunnel boring machine performance by neuro-fuzzy methods. Tunn. Undergr. Sp. Technol. **15**, 259–269 (2000)

19. N. Barton, R. Lien, J. Lunde, Engineering classification of rock masses for the design of tunnel support. Rock Mech. **6**, 189–236 (1974)

20. Z.T. Bieniawski, B. Celada, J.M. Galera, M. Álvares, Rock mass excavability (RME) index, in *ITA World Tunneling Congress Korea* (2006)

21. A. Palmstrom, E. Broch, Use and misuse of rock mass classification systems with particular reference to the Q-system. Tunn. Undergr. Sp. Technol. **21**, 575–593 (2006)

22. B. Singh, R.K. Goel, *Engineering Rock Mass Classification: Tunneling, Foundations, and Landslides, Waltham* (Butterworth-Heinemann, MA, 2011)

23. B. Maidl, L. Schmid, W. Ritz, M. Herrenknecht, *Hardrock Tunnel Boring Machines* (John Wiley & Sons, 2008)

24. S. Okubo, K. Fukui, W. Chen, Expert system for applicability of tunnel boring machines in Japan. Rock Mech. Rock Eng. **36**, 305–322 (2003)

25. B. Stack, *Handbook of Mining and Tunnelling Machinery* (Chichester, Wiley, 1982), p. 776. in Int. J. Rock Mech. Min. Sci. Geomech. Abstr. (Pergamon, 1982) p. 137

26. A. Pelzer, Die Entwicklung der Streckenvortriebsmaschinen im In-und Ausland. Glückauf. **90**, 1648–1658 (1954)

27. D.A. für unterirdisches Bauen, Empfehlungen zur Auswahl und Bewertung von Tunnelvortriebsmaschinen, Tunnel. **5**, 20–35 (1997)

28. U. Beckmann, Tunnel-boring machine payment on basis of actual rock quality effect (1982)

29. B. Singh, R.K. Goel, *Rock Mass Classification: A Practical Approach in Civil Engineering* (Elsevier, 1999)

30. D.J. Armaghani, M. Koopialipoor, A. Marto, S. Yagiz, Application of several optimization techniques for estimating TBM advance rate in granitic rocks. J. Rock Mech. Geotech. Eng. (2019). https://doi.org/10.1016/j.jrmge.2019.01.002

31. T. Kim, Development of a fuzzy logic based utilization predictor model for hard rock tunnel boring machines (2004)

32. R.A. Snowdon, M.D. Ryley, J. Temporal, A study of disc cutting in selected British rocks, in Int. J. Rock Mech. Min. Sci. Geomech. Abstr. 107–121 (1982). (Elsevier)

33. H. Yang, J. Liu, B. Liu, Investigation on the cracking character of jointed rock mass beneath TBM disc cutter. Rock Mech. Rock Eng. **51**, 1263–1277 (2018)

34. M.A. Grima, P.N.W. Verhoef, Forecasting rock trencher performance using fuzzy logic. Int. J. Rock Mech. Min. Sci. **36**, 413–432 (1999)

35. J. Zhou, B. Yazdani Bejarbaneh, D. Jahed Armaghani, M.M. Tahir, Forecasting of TBM advance rate in hard rock condition based on artificial neural network and genetic programming techniques. Bull. Eng. Geol. Environ. **79**, 2069–2084 (2020). https://doi.org/10.1007/s10064-019-01626-8

36. J. Zhou, Y. Qiu, S. Zhu, D.J. Armaghani, C. Li, H. Nguyen, S. Yagiz, Optimization of support vector machine through the use of metaheuristic algorithms in forecasting TBM advance rate. Eng. Appl. Artif. Intell. **97**(n.d.), 104015 (2021). https://doi.org/10.1016/j.engappai.2020.104015

37. M. Koopialipoor, S.S. Nikouei, A. Marto, A. Fahimifar, D.J. Armaghani, E.T. Mohamad, Predicting tunnel boring machine performance through a new model based on the group method of data handling. Bull. Eng. Geol. Environ. **78**, 3799–3813 (2018)

38. J. Zhou, Y. Qiu, D.J. Armaghani, W. Zhang, C. Li, S. Zhu, R. Tarinejad, Predicting TBM penetration rate in hard rock condition: a comparative study among six XGB-based metaheuristic techniques. Geosci. Front. (2020). https://doi.org/10.1016/j.gsf.2020.09.020

39. P. Bruines, Neuro-fuzzy modeling of TBM performance with emphasis on the penetration rate. Mem. Cent. Eng. Geol. Netherlands Delft. **202** (1998)

40. N. Innaurato, A. Mancini, E. Rondena, A. Zaninetti, Forecasting and effective TBM performances in a rapid excavation of a tunnel in Italy, in *7th ISRM Congress, International Society for Rock Mechanics and Rock Engineering* (1991)

41. R. Ribacchi, A.L. Fazio, Influence of rock mass parameters on the performance of a TBM in a gneissic formation (Varzo Tunnel). Rock Mech. Rock Eng. **38**, 105–127 (2005)

42. A. Bruland, *Hard Rock Tunnel Boring* (Norwegian University of Science and Technology, Trondheim, 1998)

43. A. Lislerud, Hard rock tunnel boring: prognosis and costs. Tunn. Undergr. Sp. Technol. **3**, 9–17 (1988)

44. A. Bruland, Hard rock tunnel boring advance rate and cutter wear. Trondheim Nor. Inst. Technol. (1999)

45. A. Palmstrom, RMi-a rock mass characterization system for rock engineering purposes (1995)

46. Z.T. Bieniawski, Predicting TBM excavability. Tunnels Tunn. Int. (2007)

47. Z.T. Bieniawski, R. Grandori, Predicting TBM excavability-part II. Tunn. Tunn. Int. **25** (2007)

48. N. Barton, Z.T. Bieniawski, RMR and Q-Setting Record Straight. Tunn. Tunn. Int. (2008)

49. R.K. Goel, Evaluation of TBM performance in a Himalayan tunnel, in *Proceedings of World Tunnel Congress* (India, 2008), pp. 1522–1532

50. J. Hassanpour, J. Rostami, M. Khamehchiyan, A. Bruland, H.R. Tavakoli, TBM performance analysis in pyroclastic rocks: a case history of Karaj water conveyance tunnel. Rock Mech. Rock Eng. **43**, 427–445 (2010)

51. K. Oraee, B. Salehi, Assessing prediction models of advance rate in tunnel boring machines—a case study in Iran. Arab. J. Geosci. **6**, 481–489 (2013)

52. D.J. Armaghani, *Tunnel Boring Machine Performance Prediction in Tropically Weathered Granite Through Empirical and Computational Methods* (Universiti Teknologi Malaysia, Johor, Malaysia, 2015)

53. W. Ritter, Die statik der tunnelgewölbe, J. (Springer, 1879)

54. D.U. Deere, A.J. Hendron, F.D. Patton, E.J. Cording, Design of surface and near-surface construction in rock, in *8th US Symposium Rock Mechanics, American Rock Mechanics Association* (1966)
55. Z.T. Bieniawski, Engineering classification of jointed rock masses. Civ. Eng. South Africa. **15** (1973)
56. Z.T. Bieniawski, Engineering rock mass classifications: a complete manual for engineers and geologists in mining, civil, and petroleum engineering. (John Wiley & Sons, 1989)
57. G.E. Wickham, Hr. Tiedemann, E.H. Skinner, Support determinations based on geologic predictions, in *N Am Rapid Excavation Tunnel Conference Proceeding* (1972)
58. E. Hoek, E.T. Brown, Practical estimates of rock mass strength. Int. J. Rock Mech. Min. Sci. **34**, 1165–1186 (1997)
59. E.M. Tallon, Comparison and application of geomechanics classification schemes, in *Tunnel Construction: Tunnelling 82, Proceedings of the 3rd International Symposium*, Brighton, 7–11 June 1982, P241–246 (Publ London, IMM, 1982), in Int. J. Rock Mech. Min. Sci. Geomech. Abstr., (Pergamon 1983), p. A10
60. F. Cassinelli, S. Cina, N. Innaurato, Power consumption and metal wear in tunnel-boring machines: analysis of tunnel-boring operation in hard rock, in *Tunneling 82, Proceedings of the 3rd International Symposium*, Brighton, 7–11 June 1982, P73–81 (Publ London, IMM, 1982), in Int. J. Rock Mech. Min. Sci. Geomech. Abstr., (Pergamon, 1983), p. A25
61. Y. Abd Al-Jalil, P.P. Nelson, C. Laughton, TBM performance analysis and rock mass impacts. Int. Symp. Mine Mech. Autom. 201–209 (1993)
62. E. Tonnizam Mohamad, D. Jahed Armaghani, M. Ghoroqi, B. Yazdani Bejarbaneh, T. Ghahremanians, M.Z. Abd Majid, O. Tabrizi, Ripping production prediction in different weathering zones according to field data. Geotech. Geol. Eng. **35** (2017). https://doi.org/10.1007/s10706-017-0254-4
63. A. Mahdiyar, M. Hasanipanah, D.J. Armaghani, B. Gordan, A. Abdullah, H. Arab, M.Z.A. Majid, A Monte carlo technique in safety assessment of slope under seismic condition. Eng. Comput. **33**, 807–817 (2017). https://doi.org/10.1007/s00366-016-0499-1
64. D. Jahed Armaghani, M.F. Mohd Amin, S. Yagiz, R.S. Faradonbeh, R.A. Abdullah, Prediction of the uniaxial compressive strength of sandstone using various modeling techniques. Int. J. Rock Mech. Min. Sci. **85**, 174–186 (2016). https://doi.org/10.1016/j.ijrmms.2016.03.018
65. D.J. Armaghani, E.T. Mohamad, M. Hajihassani, S. Yagiz, H. Motaghedi, Application of several non-linear prediction tools for estimating uniaxial compressive strength of granitic rocks and comparison of their performances. Eng. Comput. **32**, 189–206 (2016)

Chapter 2
Empirical, Statistical, and Intelligent Techniques for TBM Performance Prediction

Abstract The use of tunnel boring machine (TBM) in mechanized tunnelling excavation in various ground conditions has been highlighted in many projects. In these projects, estimation of the TBM performance is considered as a significant issue since it can be an influential parameter related to the project cost. Hence, many scholars tried to develop simple, applicable, and powerful methodologies for the prediction of TBM performance. The total developed methods in this regard can be divided into four categories, namely, theoretical, empirical, statistical, and computational. In this study, the advantages and disadvantages of these techniques were discussed. Many investigators mentioned that empirical and theoretical techniques are not good enough in accurate prediction of TBM performance. Some other researchers developed statistical-based models/equations in predicting TBM performance. However, their accuracy level is only suitable (coefficient of determination ~0.6) in many cases. On the other hand, these techniques are not good if there are some outlier data samples in the database. The best model category for TBM performance prediction is related to machine learning (ML) and artificial intelligence (AI) techniques. Using these techniques, a complex problem (i.e., TBM performance) can be solved with a high level of accuracy and low level of system error (coefficient of determination ~0.9). This study concluded that ML and AI are considered as accurate, powerful, and simple techniques in the area of tunnelling and they can be used in other applications of geotechnics as well.

Keywords Tunnel boring machine · TBM performance · Theoretical and empirical techniques · Machine learning · Statistical-Based techniques

2.1 Introduction

From the first development of the tunnel boring machine (TBM) in 1954, many improvements have been carried out on the design of this machine to increase its performance capacity. All the developments mentioned above resulted in more powerful TBMs that can be successfully applied in different ground conditions, from very hard grounds to soft ones. Predicting TBM performance in difficult rock mass

© The Author(s), under exclusive license to Springer Nature Singapore Pte Ltd. 2021   17
D. Jahed Armaghani and A. Azizi, *Applications of Artificial Intelligence in Tunnelling and Underground Space Technology*, SpringerBriefs in Applied Sciences and Technology, https://doi.org/10.1007/978-981-16-1034-9_2

conditions has posed a challenge in operations [1, 2]. In a tunnelling project, geological condition, mass weathering, strength, etc. affect TBM performance. Thus, the prediction of TBM performance is of vital importance when tunnel projects are being planned [3]. It also contributes to the selection of suitable construction methods as well as the reduction of risks connected with high capital costs; a very common issue in mechanized excavation operations [4].

Currently, TBMs are applied to massive scale of tunnelling and underground construction in both mining industry and civil construction. Along with the recent development of machine manufacturing and rapid advancements occurred in technology, in particular information and computer technologies, influential and somewhat smart machines have appeared in the market. Effective TBM planning is dependent upon the precise estimation of TBM performance such as the penetration rate (PR) and the advance rate (AR). In recent decades, several studies have been conducted to come up with models that can predict TBM performance more accurately and comprehensively. Most early models did not take into account the impact of rock mass discontinuities despite several scholars' assertion that discontinuities within a rock mass can affect TBM performance. Since these rock mass properties are subjected to many factors such as climate, rock type, weathering state, and geomorphology, there is a need to develop TBM performance predictive model for the respective region [5].

Many methods and classifications have been proposed for the prediction of TBM performance through machine factors and multiple rock factors (i.e., material and mass). In these methods and classifications, researchers have referred to several parameters as predictors of TBM performance. These parameters are considered within three groups (1) machine factors, (2) material properties of the rock, and (3) mass properties of the rock.

In this chapter, all developed models in the area of TBM performance prediction were divided into four groups, i.e., (1) theoretical, (2) empirical, (3) statistical, and (4) computational (which is based on machine learning and soft computing techniques). The background and relevant conducted studies in each group will be described in this study. Then, the advantages and disadvantages of these methods will be discussed and the best methodologies will be introduced for practical use in the area of tunnelling projects using TBM.

2.2   Theoretical Models

Researchers have adopted different approaches to the evaluation of TBM performance prediction. Some of the most important theoretical methods are discussed as follows.

2.2.1 Cutter Load Approach

To find force equilibrium equations, on the basis of the failure mechanism of the rock, the theoretical models make an analysis on cutting forces that act on the disc cutter [6–9]. The theoretical models were developed primarily through the use of full-scale tests of laboratory cutting or indentation tests to make an estimation of the cutting forces considering the cutting geometry, cutter, spacing, and penetration of the cutter. Cutter head revolution per minute (RPM), installed power, disc spacing, and thrust are key parameters in the TBM design, which have an effect on the resulting PR. RPM and machine thrust are determined by disc rolling velocity and cutter head loading capacity, respectively.

Roxborough and Phillips [10] applied basic principles and cutting geometry to calculate the theoretical normal and rolling forces on a single V-shaped disc cutter. A formula was proposed to obtain the normal and rolling forces from the uniaxial compressive strength (UCS) of the rock, disc diameter, and penetration.

$$F_N = 4 . UCS . tan\frac{\phi}{2}\sqrt{d . P^3 - P^4} \tag{2.1}$$

where, P = penetration, d = diameter of the disc, F_N = normal force, ϕ = one-half of cutter tip angle. Therefore, the following formula is related to the rolling force estimation:

$$F_R = 4 . UCS . P^2 . Tan\,\phi/2 \tag{2.2}$$

Sanio [7] suggested the following formula to predict cutting forces:

$$F_N = 2 . p . Tan\left(\frac{\theta}{2}\right)\sigma_0 \tag{2.3}$$

where θ equals the tip wedge angle and σ_0 denotes the hydrostatic pressure in the crushed zone. Sanio's work was followed by Sato et al. [11] and they introduced the following formula in this regard:

$$F = k . P^a . S^b \tag{2.4}$$

where k = coefficient of cutting, a = penetration coefficient, F = Force, P = penetration, S = cutter spacing, b = spacing coefficient.

Colorado School of Mines (CSM) developed a model, which has been named after the same school. Ozdemir [12] developed the first version of the model and Rostami [8] updated it. In the CSM model, the cutter forces for a given penetration (mm/rev) are estimated according to rock properties as well as the cutter and cutting geometry.

The equations proposed by Rostami [13] have been applied quite successfully in different projects. The total estimated resultant cutting force is shown in the following formula:

$$F_t = \int_0^\phi TRPd\theta = \int_0^\phi TRP^\circ \left(\frac{\theta}{\phi}\right)^\Psi d\theta = \frac{TRP^\circ \phi}{1 + \Psi} \qquad (2.5)$$

where T = cutter tip width, F_t = total resultant force, ϕ = angle of contact area between rock and cutter, and R = cutter radius.

$$P_c = P^o \left(\frac{\alpha}{\phi}\right)^\Psi \text{ and } \phi = Cos^{-1}\left(\frac{R - p_c}{R}\right) \qquad (2.6)$$

where Ψ = power of pressure function, P_c = pressure of crushed zone, P^o = base pressure in the crushed zone at the point directly underneath the cutter, α = position angle.

The CSM model suffered from a drawback; it did not consider quantitatively the properties of rock mass, including rock brittleness, fracture orientations, and planes of weakness. Yagiz and Ozdemir [14] and Yagiz [4] modified the CSM model by adding the indices of intact rock brittleness and rock mass fracture properties to the model. Ramezanzadeh [15] developed a TBM field performance database for more than 60 km of tunnels, and in the case of discontinuities and joints, adjustment factors were offered for the CSM models.

2.2.2 Specific Energy Approach

The relationship between normal rolling forces and penetration was introduced by Snowdon et al. [6] after a comprehensive study of disc cutting in British rocks. The assertion was made that there is a critical penetration for each spacing and rock type combination, beyond which further reduction of specific cutting energy cannot be achieved. It was also shown that the forces increase almost linearly with spacing until the S/P value of 15–20 is reached. Using selected British rocks, Snowdon et al. [6] found a relationship between the normal and rolling forces and defined it as follows:

$$\frac{F_{Normal}}{F_{Rolling}} = 21.71 \cdot p^{-0.656} \qquad (2.7)$$

where p is penetration per revolution.

Adopting a totally different approach, Boyd [16] developed a novel model. In his model, the assumption was made that specific energy is required to disintegrate the rock mass. When the tunnel cross-sectional area and the installed cutter head power are known, the following equation can be applied to estimate the PR:

$$PR = \frac{HP.\eta}{SE.A} \tag{2.8}$$

where HP is the installed cutter head power (kW), A is tunnel cross-sectional area (m^2), PR is the penetration rate (m/h), η is the mechanical efficiency factor, and SE is specific energy (kWh/m^3).

The theoretical models have been developed mainly by the use of full-scale laboratory cutting tests or indentation tests. The most important drawback of laboratory cutting tests is the fact that they do not show entirely the conditions of the actual rock mass TBM disc cutters encounter in the field [17]. In addition, the equipment of this type of tests might not be available in all of the research centres in the world [17]. When it is not possible to perform such tests, TBM performance predictions can be conducted by adjusting performance data taken from sites where a rock with similar strength properties was bored. Additionally, according to Farrokh et al. [18], having several parameters is another disadvantage of theoretical models.

2.3 Empirical Models

Several studies have been carried out for correlating the results of laboratory index test to the TBM performance. Prediction equations can be developed empirically or derived from force equilibrium or energy balance theories. In the mentioned techniques, coefficients are derived based on correlations among some given parameters in the database to simplify disc indentation geometry and the distribution of contact zone stress.

In the model introduced by Graham [19], the PR was calculated as one function of the typical forces per cutter the RPM, and UCS of rock between 140 MPa and 200 MPa. In this model, the discontinuities and cutter properties were not taken into account.

$$P_{rev} = \frac{3940 \times F_n}{\sigma_{cf}} \tag{2.9}$$

where σ_{cf} is uniaxial compressive strength (kN/m^2), F_n is average cutter force (kN), and P_{rev} is penetration per revolution (mm/rev).

Considering the results of the average cutter force and the rock tensile strength, Farmer and Glossop [20] developed a model for TBM PR prediction. This model was on the basis of eight case histories, which can be the most important limitation in regard to a wide variety of available TBMs. This model did not take into consideration the cutter geometry and rock mass properties (i.e., discontinuity).

$$P_{rev} = \frac{624 \times F_n}{\sigma_t} \tag{2.10}$$

where σ_t is tensile strength (kN/m^2), F_n is average cutter force (kN), and P_{rev} is penetration per revolution (mm/rev).

Similar to the model proposed by Graham [19], Hughes [21] developed another model in which the force per cutter, RPM, and UCS of the rock were taken into consideration. In addition, this model included the radius of the discs and the number of cutters per kerf (groove). On the other hand, rock discontinuities were not considered in this model. Hughes [21] made a prediction regarding the power requirement and performance rate of full-face machines that were provided with disc in coal measure strata. In this formula, speed of cutting, thrust per disc, average number of discs per kerf, UCS of intact rock, and average radius of discs were incorporated. The equation developed is as follows:

$$PR = \frac{6 \times P_d^{1.2} \times N \times h}{\sigma_{cf}^{1.2} r^{0.6}} \tag{2.11}$$

where PR is the rate of penetration (m/h), P_d is thrust per disc periphery (kN), N is cutting head speed (rev/s), h is the average number of discs per kerf, σ_{cf} is the compressive strength of intact rock (MPa), and r is average disc radius (m).

As a conclusion of this part, some advantages and disadvantages of the mentioned methods are discussed. It can be concluded that laboratory models/equations are considered easy to apply. This is because of their limited number of effective variables for solving TBM performance-related issues. However, in the mentioned models/equations, many of the important factors affecting TBM performance such as relevant rock properties are not considered and calculated [18]. According to Ramezanzadeh [15], a single rock property (UCS) and machine property are not enough to predict the performance of TBM. As mentioned by many researchers, other parameters, such as joint condition, tensile strength, have significant effects on TBM performance [22–24].

2.4 Statistical Approach

Statistical methods have been extensively utilized to predict TBM performance. The purpose of using and developing these techniques is to find a function/relation between predictors and model outputs. Typically, we have linear multiple regression (LMR) and non-linear multiple regression (NLMR) techniques for solving science and engineering problems. Through LMR and NLMR techniques, a linear and non-linear equation/model can be found for predicting model output using model inputs. In the following paragraphs, some of such proposed models in predicting TBM performance are described.

Hassanpour et al. [13] used data obtained from Manapouri tunnel project in New Zealand and three TBM projects in Iran to propose a new classification for boreability as well as a TBM performance model. Using regression analysis, an equation was developed by Hassanpour et al. [13] to predict FPI values as follows:

$$FPI = \exp\left(0.008\,UCS + 0.015\,RQD + 1.384\right) \tag{2.12}$$

where FPI, UCS, and RQD are field penetration index, uniaxial compressive strength (MPa), and rock quality designation (%), respectively.

A NLMR model was suggested to predict PR in the study carried out by Oraee et al. [25]. For this purpose, they used 177 datasets including three input parameters (i.e., RQD, DPW, and UCS) and one output (PR), which were obtained from two tunnel projects namely Queens Water Tunnel, USA and Gilgel Gibe II hydroelectric project, Ethiopia. The coefficient of determination (R^2) of 0.42 between measured and predicted PR values was recommended in their study. Farrokh et al. [18] carried out a comprehensive review of more than 300 TBM projects to develop a new model for predicting PR. A new power equation was developed by Farrokh et al. [18] ($R^2 = 0.58$) to predict PR as follows:

$$PR = \frac{Fn^{0.186}\,.\,RQD_c^{0.133}\,.\,RT_c^{0.183}\,.\,RPM^{0.363}\,.\,D^{5.47}\,.\,exp\left(0.046\,.\,D^2\right)}{5.64\,.\,UCS^{0.248}\,.\,exp(1.58\,.\,D)} \tag{2.13}$$

where Fn is disc cutter normal force (kN), $RQDc$ is RQD code, D is tunnel diameter (m), and RTc is rock type code.

Delisio et al. [26] proposed FPI_{blocky} equation based on volumetric joint count (J_v) and UCS data obtained from Lötschberg Base Tunnel, Switzerland in order to predict TBM performance in blocky rock mass as follows:

$$FPI_{blocky} = 5952 - 1794 \ln J_v + UCS \tag{2.14}$$

Subsequently, they introduced PR_{blocky} and Net AR_{blocky} as follows:

$$PR_{blocky} = Thrust\ Force/FPI_{blocky} \tag{2.15}$$

$$\text{Net } AR_{blocky} = PR_{blocky} \times 60 \times RPM/1000 \tag{2.16}$$

where, PR_{blocky} signifies penetration rate in blocky rock mass and Net AR_{blocky} represents the net advance rate in blocky rock mass. Finally, total AR_{blocky} was formulated as:

$$\text{Total } AR_{blocky} = Net\ AR_{blocky} \times UF \times 24 \tag{2.17}$$

where UF is the TBM utilisation factor (%).

In another study of regression analysis, Mahdevari et al. [27] proposed new multiple equations for the prediction of PR using 150 data points obtained from Queens Water Tunnel, USA. Their proposed equation using LMR model is shown as follows:

$$PR = 2.0556 - (0.0024 \times UCS) - (0.0098 \times BTS) + (0.0039 \times BI) - (0.0296 \times DPW) - (0.0052 \times \alpha) - (0.1091 \times SE) - (0.0004 \times TF) - (0.1154 \times CP) + (0.1027 \times CT)$$

$$(2.18)$$

R^2 of 0.771 shows this equation is able to predict PR with relatively high accuracy.

Farrokh et al. [18] showed that compared to simple models, multiple parameter models use more project-specific data and are easier to use in comparison with computer-aided and artificial intelligence (AI) models.

2.5 Computational-Based Techniques

In general, AI and machine learning (ML) methods as powerful tools are one of the most dynamic research fields in advanced and diverse applications of both science and engineering [28–52]. The most interesting feature of this approach is that subjectivity and uncertainty are treated scientifically through an engineering process, rather than blindly avoided [3]. The capability of these techniques in the field of geotechnical engineering, and more specifically, in underground space technologies, has been highlighted by many researchers in which some of them can be seen in the following paragraphs.

Benardos and Kaliampakos [53] examined 1077 m length of Athens Metro tunnel to develop an artificial neural network (ANN) model for the prediction of AR. They used several parameters, i.e., RQD, RMR, UCS, groundwater table, overburden, overload factor, and permeability as input parameters to predict AR. They pointed out that the developed ANN can be utilized to predict TBM AR and identify the risk-prone areas as a practical tool. Based on the data from three TBM projects, Simoes and Kim [54] applied two fuzzy inference system (FIS) types called rule-based and parametric-based systems to predict the utilization index (UI) of TBM. They showed a high capability of FIS technique in predicting TBM UI. Javad and Narges [55] developed an ANN model for estimating TBM PR using an appropriate 185 data samples. The R^2 for the developed ANN model was obtained as 0.94. Using 7.5 km data of Queens Water Tunnel in the USA, Yagiz and Karahan [56] introduced a particle swarm optimization (PSO) model for the prediction of PR. Considering the results of UCS, BTS, BI, DPW, and α, they developed an equation with correlation coefficient (R) values of 0.821 and 0.737 for training and testing datasets, respectively. Extreme learning machine (ELM) technique as a specific type of ANN was used to forecast TBM PR in the study carried out by Shao et al. [57]. They concluded that the proposed model can predict PR better in comparison to other mentioned methods such as support vector machine.

Two intelligent equations were developed for the prediction of PR and AR, respectively using gene expression programming (GEP) and genetic programming (GP) techniques in the studies by Armaghani et al. [23] and Zhou et al. [58]. These researchers concluded that their proposed GEP and GP models are considered as a practical tool/solution in the area of TBM performance. In a group of TBM performance studies, a series of optimization techniques such as Moth Flame Optimization (MFO), grey wolf optimizer (GWO), imperialism competitive algorithm (ICA) were used to optimize the basic ML and AI techniques such ANN and support vector machine (SVM) [5, 24, 59–61]. In another study of TBM performance, Koopialipoor et al. [62] used a modern version of ANN namely deep neural network (DNN) for solving PR problem and confirmed that the proposed DNN model can be applied in this area. Table 2.1 shows a summary of recent studies on TBM performance prediction using AI and ML approaches. According to this table, a higher performance prediction can be obtained by using and developing AI and ML techniques in estimating TBM performance.

AI and ML techniques provide a more complex methodology to estimate TBM performance. Models applying AI should only be used when detailed information of a similar condition is available in the tunnelling project. Substantial errors may take place applying the model if there are significant differences in ground conditions. Another common problem with AI models is that these models are rarely applied for TBM performance estimation purposes in practice, despite the fact that they have many advantages compared to other methods.

Additionally, Farrokh et al. [18] stated that AI and ML methods are not available in public domain. It should be noted that proposing a TBM performance model with higher degree of accuracy compared to previous methods is always of interest. Although AI and ML techniques have several disadvantages, they can predict TBM performance better. According to Table 2.1, generally, $R^2 \sim 0.9$ for various computing techniques reveals that AI and ML can perform successfully in the field of TBM performance prediction.

2.6 Discussion and Conclusion

The performance prediction of TBM is essential to know before starting a tunnelling project or even ordering a TBM machine. This is important especially for scheduling operation period, estimation cost of the projects and minimizing risks associated with such projects. Due to that, many researchers tried to solve this problem by developing a new methodology/idea, which can be categorized into four groups, i.e., theoretical, empirical, statistical-based, and computational-based. The theoretical models were proposed according to data from full-scale cutting tests, which can be divided into two general groups of specific energy and cutter load. The most important disadvantage

Table 2.1 Some of AI and ML works in the area of TBM performance prediction

References	Model	Output	Description	R^2
Ghasemi et al. [63]	FIS	PR	Using 151 datasets	$R^2 = 0.89$
Gholamnejad and Tayarani [55]	ANN	PR	Using 185 datasets	$R^2 = 0.94$
Benardos and Kaliampakos [53]	ANN	AR	–	–
Simoes and Kim [54]	FIS	UI	Using data of three TBM projects	–
Grima et al. [3]	ANN, ANFIS	PR, AR	A database consisting 640 TBM projects	–
Yagiz and Karahan [56]	PSO	PR	Number of 151 datasets	$R^2 = 0.67$
Mikaeil et al. [64]	FIS	PR	Using dataset presented by Yagiz [22]	–
Yagiz et al. [65]	ANN	PR	Using 151 datasets	$R^2 = 0.9$
Gholami et al. [66]	ANN	PR	Data of 121 tunnel sections	$R^2 = 0.72$
Salimi and Esmaeili [67]	ANN	PR	Data of 46 sections of the Karaj–Tehran tunnel	$R^2 = 0.83$
Torabi et al. [68]	ANN	PR, UI	Data of Tehran–Shomal highway project	$R^2_{PR} = 0.99$ $R^2_U = 0.99$
Armaghani et al. [5]	ICA-ANN	PR	Using 1286 data samples in hard rock condition	$R^2 = 0.92$
Armaghani et al. [60]	POS-ANN	AR	Using 1286 data samples in hard rock condition	$R^2 = 0.95$
Oraee et al. [25]	ANFIS	PR	Using 177 datasets obtained from two tunnel projects	$R^2 = 0.69$
Yang et al. [69]	GWO-FW-MKL-SVR	PR	Using a database comprising of 503 data samples	$R^2 = 0.94$
Mahdevari et al. [27]	SVR	PR	150 data points pertaining to the Queens Water Tunnel, USA	$R^2 = 0.98$ lePara>
Zhang et al. [70]	BO	PR	Number of 151 data samples	–
Salimi et al. [71]	SVR	FPI	Number of 75 data samples	$R^2 = 0.91$

(continued)

Table 2.1 (continued)

References	Model	Output	Description	R^2
Armaghani et al. [23]	GEP	PR	Using 1286 data samples in hard rock condition	$R^2 = 0.85$
Zhou et al. [58]	GP	AR	Using 1286 data samples in hard rock condition	$R^2 = 0.91$
Koopialipoor et al. [72]	DNN	PR	Using 1286 data samples in hard rock condition	$R^2 = 0.93$
Zhou et al. [61]	SVM-MFO	AR	Using 1286 data samples in hard rock condition	$R^2 = 0.96$
Zhou et al. [59]	PSO-XGB	PR	Using 1286 data samples in hard rock condition	$R^2 = 0.95$

Nomenclature: Adaptive Neuro Fuzzy Inference System (ANFIS); Feature Weighted (FW); Bayesian Optimization (BO); Support Vector Regression (SVR); Particle Swarm Optimization (PSO); Moth Flame Optimization (MFO); Multiple Kernel (MK); Genetic Programming (GP); Deep Neural Network (DNN); Support Vector Machine (SVM); Extreme Gradient Boosting (XGB); Gene Expression Programming (GEP); Grey Wolf Optimizer (GWO)

of such models is the fact that they do not show entirely the conditions of the actual rock mass TBM disc cutters encounter in the field. Another drawback of theoretical models is related to their low performance in predicting TBM performance. The amount of errors associated with theoretical models is dependent on the accuracy of the underlying model assumptions, and the quality and quantity of the data related to TBM and ground conditions. In addition, as mentioned by Farrokh et al. [18], these techniques need several factors to do the analysis and estimation, which is another disadvantage of them.

The second general techniques to estimate TBM performance are empirical, which are based on a single rock mass or material property such as rock strength or a combination of two or three of them. The results of these parameters should be obtained in the laboratory according to the available relevant standards. Although the laboratory tests are easy to conduct and use, they can only offer a limited range of application. Another limitation of such techniques is related to the type of model predictors. In these techniques, many of the parameters are related to rock material properties, which can be easily measured in the laboratory. The parameters related to the rock mass and condition such as RQD, RMR, and rock type were not considered in empirical techniques.

The next group of techniques in solving TBM performance is statistical-based approaches. These techniques are extensively used in different fields of engineering [73–76]. By implementing these techniques, linear or non-linear relationships/equations between model predictors and model target (i.e., TBM performance) can be introduced. These equations that are multiple are consisting of at least two model predictors to a number of parameters. However, according to Alvarez Grima and Verhoef [77], these techniques are not always robust enough to render accurate accounts of nonlinear and complex systems. Besides, their performance capacity becomes poor if the data include several outliers. Hence, in order to get the highest performance capacity in the area of TBM performance, there is a need to propose ML and AI techniques. The AI and ML techniques, because of their nature, are able to make a non-linear relation between inputs or predictors and output parameters. These techniques include several shortcomings and advantage as well. Regarding shortcomings, it should be mentioned that these models can be used when detailed information of a similar condition is available. Therefore, such models depend on performance data collected from similar case histories. Another common problem with such new methodologies is that they are rarely applied in practice not only in the area of TBM performance prediction but also in other science and engineering fields. Nevertheless, when a high level of model accuracy is of interest and advantage, AI and ML could be considered as the best choice. By using some of these techniques such as GEP and GP, a practical equation, which is intelligent at the same time, can be developed for TBM performance prediction. Such intelligent equations can be used in the site investigation phase where we should have some estimations of TBM performance, operation period, and so on.

References

1. M. Sapigni, M. Berti, E. Bethaz, A. Busillo, G. Cardone, TBM performance estimation using rock mass classifications. Int. J. Rock Mech. Min. Sci. **39**, 771–788 (2002)
2. J. Zeng, B. Roy, D. Kumar, A.S. Mohammed, D.J. Armaghani, J. Zhou, E.T. Mohamad, Proposing several hybrid PSO-extreme learning machine techniques to predict TBM performance. Eng. Comput. (2021). https://doi.org/10.1007/s00366-020-01225-2
3. M.A. Grima, P.A. Bruines, P.N.W. Verhoef, Modeling tunnel boring machine performance by neuro-fuzzy methods. Tunn. Undergr. Sp. Technol. **15**, 259–269 (2000)
4. S. Yagiz, Development of rock fracture and brittleness indices to quantify the effects of rock mass features and toughness in the CSM Model basic penetration for hard rock tunneling machines (2002)
5. D.J. Armaghani, E.T. Mohamad, M.S. Narayanasamy, N. Narita, S. Yagiz, Development of hybrid intelligent models for predicting TBM penetration rate in hard rock condition. Tunn. Undergr. Sp. Technol. **63**, 29–43 (2017). https://doi.org/10.1016/j.tust.2016.12.009
6. R.A. Snowdon, M.D. Ryley, J. Temporal, A study of disc cutting in selected British rocks. Int. J. Rock Mech. Min. Sci. Geomech. Abstr. (Elsevier, 1982), pp. 107–121
7. H.P. Sanio, Prediction of the performance of disc cutters in anisotropic rock. Int. J. Rock Mech. Min. Sci. Geomech. Abstr. (Elsevier, 1985), pp. 153–161
8. J. Rostami, Development of a force estimation model for rock fragmentation with disc cutters through theoretical modeling and physical measurement of crushed zone pressure (1997)
9. J. Rostami, L. Ozdemir, A new model for performance prediction of hard rock TBMs, in *Proceedings Rapid Excavation and Tunneling Conference, Society for Mining, Metallogy & Exploration* (Inc, 1993), p. 793
10. F.F. Roxborough, H.R. Phillips, Rock excavation by disc cutter, in Int. J. Rock Mech. Min. Sci. Geomech. Abstr. (Elsevier, 1975), pp. 361–366

11. K. Sato, F. Gong, K. Itakura, Prediction of disc cutter performance using a circular rock cutting ring. Proc. 1st Int. Mine Mech. Autom. Symp. (1991)
12. L. Ozdemir, Development of theoretical equations for predicting tunnel boreability (1977)
13. J. Hassanpour, J. Rostami, J. Zhao, A new hard rock TBM performance prediction model for project planning. Tunn. Undergr. Sp. Technol. 26, 595–603 (2011)
14. S. Yagiz, L. Ozdemir, Geotechnical parameters influencing the TBM performance in various rocks, in Program with Abstract, 44th Annual Meeting Association Engineering Geologists P79 (Saint Louis, Missouri, USA, 2001)
15. A. Ramezanzadeh, Performance analysis and development of new models for performance prediction of hard rock TBMs in rock mass (2005)
16. R.J. Boyd, Hard rock continuous mining machine: Mobile Miner MM-120, in Rock Excavation Engineering Seminar, Department Mining and Met. Eng (University of Quee ~ Lsland, 1986)
17. J.K. Hamidi, K. Shahriar, B. Rezai, J. Rostami, Performance prediction of hard rock TBM using Rock Mass Rating (RMR) system. Tunn. Undergr. Sp. Technol. 25, 333–345 (2010)
18. E. Farrokh, J. Rostami, C. Laughton, Study of various models for estimation of penetration rate of hard rock TBMs. Tunn. Undergr. Sp. Technol. 30, 110–123 (2012)
19. P.C. Graham, Rock exploration for machine manufacturers. Explor. Rock Eng. 173–180 (1976)
20. I.W. Farmer, N.H. Glossop, Mechanics of disc cutter penetration. Tunnels Tunn. 12, 22–25 (1980)
21. H.M. Hughes, The relative cuttability of coal-measures stone. Min. Sci. Technol. 3, 95–109 (1986)
22. S. Yagiz, Utilizing rock mass properties for predicting TBM performance in hard rock condition. Tunn. Undergr. Sp. Technol. 23, 326–339 (2008)
23. D.J. Armaghani, R.S. Faradonbeh, E. Momeni, A. Fahimifar, M.M. Tahir, Performance prediction of tunnel boring machine through developing a gene expression programming equation. Eng. Comput. 34, 129–141 (2018)
24. J. Zhou, Y. Qiu, S. Zhu, D.J. Armaghani, M. Khandelwal, E.T. Mohamad, Estimation of the TBM advance rate under hard rock conditions using XGBoost and Bayesian optimization. Undergr. Sp. (2020). https://doi.org/10.1016/j.undsp.2020.05.008
25. K. Oraee, M.T. Khorami, N. Hosseini, Prediction of the penetration rate of TBM using adaptive neuro fuzzy inference system (ANFIS), in Proceeding SME Annual Meeting Exhibation From Mine to Mark. Now It's Glob (Seattle, WA, USA, 2012), pp. 297–302
26. A. Delisio, J. Zhao, H.H. Einstein, Analysis and prediction of TBM performance in blocky rock conditions at the Ltschberg Base Tunnel, Tunn. Undergr. Sp. Technol. 33 (2013). https://doi.org/10.1016/j.tust.2012.06.015
27. S. Mahdevari, K. Shahriar, S. Yagiz, M.A. Shirazi, A support vector regression model for predicting tunnel boring machine penetration rates. Int. J. Rock Mech. Min. Sci. 72, 214–229 (2014)
28. D. Li, M.R. Moghaddam, M. Monjezi, D.J. Armaghani, A. Mehrdanesh, Development of a group method of data handling technique to forecast iron ore price. Appl. Sci. 10, 1–3 (2020). https://doi.org/10.3390/app10072364
29. D.J. Armaghani, E. Momeni, P.G. Asteris, Application of group method of data handling technique in assessing deformation of rock mass. Metaheuristic Comput. Appl. 1, 1–18 (2020)
30. M. Cai, M. Koopialipoor, D.J. Armaghani, B. Thai Pham, Evaluating slope deformation of earth dams due to earthquake shaking using MARS and GMDH techniques. Appl. Sci. 10, 1486 (2020)
31. J. Zhou, C. Chen, K. Du, D. Jahed Armaghani, C. Li, A new hybrid model of information entropy and unascertained measurement with different membership functions for evaluating destressability in burst-prone underground mines. Eng. Comput. (2020). https://doi.org/10.1007/s00366-020-01151-3
32. B.T. Pham, M.D. Nguyen, T. Nguyen-Thoi, L.S. Ho, M. Koopialipoor, N.K. Quoc, D.J. Armaghani, H. Van Le, A novel approach for classification of soils based on laboratory tests using Adaboost, Tree and ANN modeling. Transp. Geotech. 100508 (2020). https://doi.org/10.1016/j.trgeo.2020.100508

33. A. Dehghanbanadaki, M. Khari, S.T. Amiri, D.J. Armaghani, Estimation of ultimate bearing capacity of driven piles in c–φ soil using MLP-GWO and ANFIS-GWO models: a comparative study. Soft. Comput. (2020). https://doi.org/10.1007/s00500-020-05435-0

34. H. Harandizadeh, D.J. Armaghani, Prediction of air-overpressure induced by blasting using an ANFIS-PNN model optimized by GA. Appl. Soft Comput. **106904** (2020)

35. J. Huang, M. Koopialipoor, D.J. Armaghani, A combination of fuzzy Delphi method and hybrid ANN-based systems to forecast ground vibration resulting from blasting. Sci. Rep. **10**, 1–21 (2020)

36. D. Ramesh Murlidhar, B. Yazdani Bejarbaneh, B. Jahed Armaghani et al., Application of tree-based predictive models to forecast air overpressure induced by mine blasting. Nat. Resour. Res. (2020). https://doi.org/10.1007/s11053-020-09770-9

37. T.E. Asteris, P.G. Douvika, M.G. Karamani, C.A. Skentou, A.D. Chlichlia, K. Cavaleri, L. Daras, T. Armaghani, D.J. Zaoutis, A novel heuristic algorithm for the modeling and risk assessment of the COVID-19 pandemic phenomenon. Comput. Model. Eng. Sci. (2020). https://doi.org/10.32604/cmes.2020.013280

38. B.R. Murlidhar, D.J. Armaghani, E.T. Mohamad, Intelligence prediction of some selected environmental issues of blasting: a review. Open Constr. Build. Technol. J. **14**, 298–308 (2020). https://doi.org/10.2174/1874836802014010298

39. Z. Yu, X. Shi, J. Zhou, Y. Gou, X. Huo, J. Zhang, D.J. Armaghani, A new multikernel relevance vector machine based on the HPSOGWO algorithm for predicting and controlling blast-induced ground vibration. Eng. Comput. (2020). https://doi.org/10.1007/s00366-020-01136-2

40. D.J. Armaghani, M. Hajihassani, E.T. Mohamad, A. Marto, S.A. Noorani, Blasting-induced flyrock and ground vibration prediction through an expert artificial neural network based on particle swarm optimization. Arab. J. Geosci. **7**, 5383–5396 (2014)

41. W. Chen, P. Sarir, X.-N. Bui, H. Nguyen, M.M. Tahir, D.J. Armaghani, Neuro-genetic, neuro-imperialism and genetic programing models in predicting ultimate bearing capacity of pile. Eng. Comput. (2019). https://doi.org/10.1007/s00366-019-00752-x

42. Z. Shao, D.J. Armaghani, B.Y. Bejarbaneh, M.A. Mu'azu, E.T. Mohamad, Estimating the friction angle of black shale core specimens with hybrid-ann approaches. Measurement (2019). https://doi.org/10.1016/j.measurement.2019.06.007

43. J. Zhou, E. Li, H. Wei, C. Li, Q. Qiao, D.J. Armaghani, Random forests and cubist algorithms for predicting shear strengths of rockfill materials. Appl. Sci. **9**, 1621 (2019)

44. M. Khari, A. Dehghanbandaki, S. Motamedi, D.J. Armaghani, Computational estimation of lateral pile displacement in layered sand using experimental data. Measurement **146**, 110–118 (2019)

45. E. Momeni, A. Yarivand, M.B. Dowlatshahi, D.J. Armaghani, An efficient optimal neural network based on gravitational search algorithm in predicting the deformation of geogrid-reinforced soil structures. Transp. Geotech. 100446 (2020)

46. B.R. Murlidhar, D. Kumar, D. Jahed Armaghani, E.T. Mohamad, B. Roy, B.T. Pham, A novel intelligent ELM-BBO technique for predicting distance of mine blasting-induced flyrock. Nat. Resour. Res. (2020) https://doi.org/10.1007/s11053-020-09676-6

47. D.J. Armaghani, M. Hajihassani, H. Sohaei, E.T. Mohamad, A. Marto, H. Motaghedi, M.R. Moghaddam, Neuro-fuzzy technique to predict air-overpressure induced by blasting. Arab. J. Geosci. **8**, 10937–10950 (2015). https://doi.org/10.1007/s12517-015-1984-3

48. D. Jahed Armaghani, M. Hajihassani, M. Monjezi, E.T. Mohamad, A. Marto, M.R. Moghaddam, Application of two intelligent systems in predicting environmental impacts of quarry blasting. Arab. J. Geosci. **8** (2015). https://doi.org/10.1007/s12517-015-1908-2

49. H. Harandizadeh, D.J. Armaghani, E.T. Mohamad, Development of fuzzy-GMDH model optimized by GSA to predict rock tensile strength based on experimental datasets. Neural Comput. Appl. **32**, 14047–14067 (2020). https://doi.org/10.1007/s00521-020-04803-z

50. P.G. Asteris, M.G. Douvika, C.A. Karamani, A.D. Skentou, K. Chlichlia, L. Cavaleri, T. Daras, D.J. Armaghani, T.E. Zaoutis, A novel heuristic algorithm for the modeling and risk assessment of the covid-19 pandemic phenomenon. C. Comput. Model. Eng. Sci. **124**, 1–14 (2020). https://doi.org/10.32604/CMES.2020.013280

51. E.T. Mohamad, S.A. Noorani, D.J. Armaghani, R. Saad, Simulation of blasting induced ground vibration by using artificial neural network. Electron. J. Geotech. Eng. **17**, 2571–2584 (2012)
52. E.T. Mohamad, D.J. Armaghani, M. Hajihassani, K. Faizi, A. Marto, A simulation approach to predict blasting-induced flyrock and size of thrown rocks, Electron. J. Geotech. Eng. **18**(B), 365–374 (2013)
53. A.G. Benardos, D.C. Kaliampakos, Modelling TBM performance with artificial neural networks. Tunn. Undergr. Sp. Technol. **19**, 597–605 (2004)
54. M.G. Simoes, T. Kim, Fuzzy modeling approaches for the prediction of machine utilization in hard rock tunnel boring machines, in *Induced Applied Conference 2006, 41st IAS Annual Meeting Conference Record 2006 IEEE* (IEEE, 2006), pp. 947–954
55. G. Javad, T. Narges, Application of artificial neural networks to the prediction of tunnel boring machine penetration rate. Min. Sci. Technol. **20**, 727–733 (2010)
56. S. Yagiz, H. Karahan, Prediction of hard rock TBM penetration rate using particle swarm optimization. Int. J. Rock Mech. Min. Sci. **48**, 427–433 (2011)
57. C. Shao, X. Li, H. Su, Performance Prediction of Hard Rock TBM Based on Extreme Learning Machine, in Int. Conf. Intell. Robot. Appl. (Springer, 2013), pp. 409–416
58. J. Zhou, B. Yazdani Bejarbaneh, D. Jahed Armaghani, M.M. Tahir, Forecasting of TBM advance rate in hard rock condition based on artificial neural network and genetic programming techniques. Bull. Eng. Geol. Environ. **79**, 2069–2084 (2020). https://doi.org/10.1007/s10064-019-01626-8
59. J. Zhou, Y. Qiu, D.J. Armaghani, W. Zhang, C. Li, S. Zhu, R. Tarinejad, Predicting TBM penetration rate in hard rock condition: a comparative study among six XGB-based metaheuristic techniques. Geosci. Front. (2020). https://doi.org/10.1016/j.gsf.2020.09.020
60. D.J. Armaghani, M. Koopialipoor, A. Marto, S. Yagiz, Application of several optimization techniques for estimating TBM advance rate in granitic rocks. J. Rock Mech. Geotech. Eng. (2019). https://doi.org/10.1016/j.jrmge.2019.01.002
61. J. Zhou, Y. Qiu, S. Zhu, D.J. Armaghani, C. Li, H. Nguyen, S. Yagiz, Optimization of support vector machine through the use of metaheuristic algorithms in forecasting TBM advance rate. Eng. Appl. Artif. Intell. **97**(n.d.) 104015
62. M. Koopialipoor, S.S. Nikouei, A. Marto, A. Fahimifar, D.J. Armaghani, E.T. Mohamad, Predicting tunnel boring machine performance through a new model based on the group method of data handling. Bull. Eng. Geol. Environ. **78**, 3799–3813 (2018)
63. E. Ghasemi, S. Yagiz, M. Ataei, Predicting penetration rate of hard rock tunnel boring machine using fuzzy logic. Bull. Eng. Geol. Environ. **73**, 23–35 (2014)
64. R. Mikaeil, M.Z. Naghadehi, F. Sereshki, Multifactorial fuzzy approach to the penetrability classification of TBM in hard rock conditions. Tunn. Undergr. Sp. Technol. **24**, 500–505 (2009)
65. S. Yagiz, C. Gokceoglu, E. Sezer, S. Iplikci, Application of two non-linear prediction tools to the estimation of tunnel boring machine performance. Eng. Appl. Artif. Intell. **22**, 808–814 (2009)
66. M. Gholami, K. Shahriar, M. Sharifzadeh, J.K. Hamidi, A comparison of artificial neural network and multiple regression analysis in TBM performance prediction. in *ISRM Regional Symposium Asian Rock Mechanics Symposium International Society for Rock Mechanics* (2012)
67. A. Salimi, M. Esmaeili, Utilising of linear and non-linear prediction tools for evaluation of penetration rate of tunnel boring machine in hard rock condition. Int. J. Min. Miner. Eng. **4**, 249–264 (2013)
68. S.R. Torabi, H. Shirazi, H. Hajali, M. Monjezi, Study of the influence of geotechnical parameters on the TBM performance in Tehran-Shomal highway project using ANN and SPSS. Arab. J. Geosci. **6**, 1215–1227 (2013)
69. H. Yang, Z. Wang, K. Song, A new hybrid grey wolf optimizer-feature weighted-multiple kernel-support vector regression technique to predict TBM performance. Eng. Comput. (2020). https://doi.org/10.1007/s00366-020-01217-2
70. Q. Zhang, W. Hu, Z. Liu, J. Tan, TBM performance prediction with Bayesian optimization and automated machine learning. Tunn. Undergr. Sp. Technol. **103**, 103493 (2020)

71. A. Salimi, J. Rostami, C. Moormann, A. Delisio, Application of non-linear regression analysis and artificial intelligence algorithms for performance prediction of hard rock TBMs. Tunn. Undergr. Sp. Technol. **58**, 236–246 (2016)
72. M. Koopialipoor, H. Tootoonchi, D. Jahed Armaghani, E. Tonnizam Mohamad, A. Hedayat, Application of deep neural networks in predicting the penetration rate of tunnel boring machines. Bull. Eng. Geol. Environ. (2019). https://doi.org/10.1007/s10064-019-01538-7
73. A. Mahdiyar, M. Hasanipanah, D.J. Armaghani, B. Gordan, A. Abdullah, H. Arab, M.Z.A. Majid, A monte carlo technique in safety assessment of slope under seismic condition. Eng. Comput. **33**, 807–817 (2017). https://doi.org/10.1007/s00366-016-0499-1
74. E.T. Mohamad, D.J. Armaghani, A. Mahdyar, I. Komoo, K.A. Kassim, A. Abdullah, M.Z.A. Majid, Utilizing regression models to find functions for determining ripping production based on laboratory tests. Measurement **111**, 216–225 (2017)
75. E. Tonnizam Mohamad, D. Jahed Armaghani, M. Ghoroqi, B. Yazdani Bejarbaneh, T. Ghahremanians, M.Z. Abd Majid, O. Tabrizi, Ripping production prediction in different weathering zones according to field data. Geotech. Geol. Eng. **35** (2017). https://doi.org/10.1007/s10706-017-0254-4
76. B. Gordan, D.J. Armaghani, A.B. Adnan, A.S.A. Rashid, A new model for determining slope stability based on seismic motion performance. Soil Mech. Found. Eng. **53**, 344–351 (2016). https://doi.org/10.1007/s11204-016-9409-1
77. M.A. Grima, P.N.W. Verhoef, Forecasting rock trencher performance using fuzzy logic. Int. J. Rock Mech. Min. Sci. **36**, 413–432 (1999)

Chapter 3
Developing Statistical Models for Solving Tunnel Boring Machine Performance Problem

Abstract The efficiency of tunnel boring machines (TBMs) in tunnelling projects has great importance for civil and geotechnical industries. A reliable and applicable model for predicting TBM performance is of interest and necessity in any tunnelling project before construction and even ordering TBM machine. In this study, a series of statistical-based models/equations, i.e., simple regression, linear, and non-linear multiple regression (LMR and NLMR) models were developed to predict TBM performance including advance rate, AR, and penetration rate, PR. The most effective parameters on TBM performance based on different categories of rock material, rock mass, and machine properties were selected and used. Results obtained by simple regression models showed that they are not good enough for receiving a suitable accuracy in predicting TBM PR/AR. In addition, LMR and NLMR equations received a higher performance prediction compared to simple regression models. A coefficient of determination of about 0.6 confirmed a suitable and applicable accuracy level for the developed LMR and NLMR equations in estimating TBM PR/AR.

Keywords TBM performance · Simple regression · LMR · NLMR · Prediction, tunnelling project

3.1 Introduction

Over the years, rock mass classifications and field approaches have been developed in mining and civil engineering for simplifying the understanding of the complexity associated with underground spaces. To predict TBM performance, several attempts have been made to assess the capability of these approaches. These classifications and field approaches have been proposed mainly based on only rock (mass and material) properties [1–4]. Machine specifications have not been considered as predictor(s) in these systems where they play a vital role in predicting TBM performance as mentioned by several researchers [5–12]. Some of the previous studies indicated that these classifications and field approaches could not perform well in predicting TBM performance [13–15]. This is due to the site-specific behaviour of the rock mass and

© The Author(s), under exclusive license to Springer Nature Singapore Pte Ltd. 2021
D. Jahed Armaghani and A. Azizi, *Applications of Artificial Intelligence in Tunnelling and Underground Space Technology*, SpringerBriefs in Applied Sciences and Technology, https://doi.org/10.1007/978-981-16-1034-9_3

its variation from place to place [16–21]. Note that, predicting exact value of the TBM performance with high capacity is essential in planning the tunnel projects.

The feasibility and capability of multiple regression techniques for solving problems in civil engineering have been highlighted in many studies [22–35]. The objective of such methods is to produce a relationship between independent (predictor) and dependent (output) variables [36]. In case of TBM performance, several studies reported the successful application of multiple regression techniques. Based on previous investigations, they can predict TBM performance with suitable degree of accuracy in many cases [10, 37, 38].

In this study, both linear and non-linear multiple regression techniques are applied to forecast the advance rate (AR) and penetration rate (PR) of a tunnel constructed in granite rock mass in Malaysia. Considering these techniques and using the most influential parameters on TBM performance, several new equations are proposed to predict TBM performance, i.e., PR and AR. The proposed equations will be evaluated based on their prediction capacity and the best model among them will be introduced, eventually.

3.2 Regression-Based Models

3.2.1 Linear Multiple Regression (LMR)

The aim of linear multiple regression (LMR) analysis is to determine the parameters of a linear function in such a way that they function best fits a given set of observation data [33, 39–41]. In cases when at least two input variables are available, LMR can be applied to obtain the best-fit equation. Multiple regression analyses are meant to solve engineering problems through least-squares fit, which makes and solves simultaneous equations by creating a regression matrix. Employing these techniques would result in some coefficients through the backslash operator [27, 36, 42]. The extent to which a dependent variable is affected by the independent variables can be measured by regression analysis. In simple bivariate regression where only a single independent variable exists, prediction of the dependent variable can take place by considering the independent variable through the following equation:

$$y = a + bx \tag{3.1}$$

where a is constant, and y and x are dependent and independent variables, respectively. The following multiple variables can be obtained through the extension of the above equation:

$$y = a + b_1x_1 + b_2x_2 + b_3x_3 + \ldots + b_nx_n \tag{3.2}$$

where x_1, x_2, x_3, ..., x_n are different independent variables that predict y.

Several studies in different fields of geotechnical engineering have adopted the LMR technique to solve their problems [43, 44]. Grima and Babuska [43], Gokceoglu and Zorlu [44], and Jahed Armaghani et al. [45] used this method to develop a linear multiple equations for the prediction of the rock material strength. Bahrami et al. [46] suggested an LMR equation for predicting rock fragmentation caused by the blasting operations. In a study performed by Jahed Armaghani et al. [47], shear strength parameters (c and ϕ) of the rock material were estimated through the LMR technique. Hajihassani et al. [48] also proposed an equation for predicting the amount of ground vibration caused by quarry blasting, using the LMR technique.

3.2.2   Non-linear Multiple Regression (NLMR)

The NLMR model can be used in the area of science and engineering in order to solve non-linear problems. Normally, non-linear problems are more complex compared to linear ones. In NLMR, in addition to simple regression type, some other equation types such as power and logarithmic are used to find a non-linear equation [35, 49]. A common term used for the regression line is the least-squares line. In both linear and non-linear relationships such as exponential and power relationships, the NLMR technique can be used. The NLMR approach can be applied to develop mathematical formulas that make predictions about dependent variables according to known independent variables. This technique was employed by Yagiz et al. [10] and Gong and Zhao [37] to forecast TBM PR. Several NLMR equations for solving problems in the area of rock engineering and geotechnical applications were also proposed by different scholars [17, 19, 50, 51].

3.3   Case Study

The tunnelling project considered in this study is located in a central area of peninsular Malaysia to convey raw water from two states (i.e., Pahang to Selangor). The collected water is later distributed among receiving basins by pipelines through gravity flow as well as applying a planned treatment plant. The pipelines to the treatment plant and outlet connecting basin are not incorporated into the project; however, they are included in the treatment works.

The tunnel was excavated to cross the Main Range that is stretched between Pahang and Selangor States. This mountain range serves as the backbone of Peninsular Malaysia with peaks ranging from 100 to 1,400 m. To select the construction methods (i.e., TBM and conventional blasting and drilling) in this project, some factors such as economic aspect, excavation for more than 3 km, low numbers of faults, and good rock condition were considered. Finally, the tunnel excavation was

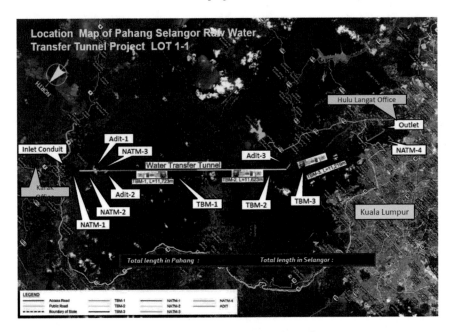

Fig. 3.1 Different construction sections in the tunnelling project [52]

carried out by three TBMs for around 35 km. The lower and upper ends were excavated using the conventional blasting and drilling technique with a total length of about 9 km. Figure 3.1 displays different construction sections in the mentioned tunnelling project. Three hard rock TBMs were designed for 34.74 km of the whole 44.6 km of the tunnel route. These TBMs were used to excavate different ground conditions, i.e., mixed ground (11,761 m), very hard ground (11,761 m), and blocky ground (11,218 m).

3.4 Data Measurement and Input Variables

As mentioned by Alvare Grima et al. [9] and Armaghani et al. [53], the most influential parameters on TBM performance can be categorized into three groups; (1) rock material properties, (2) rock mass properties, and (3) machine characteristics (i.e., TBM condition and parameters). In this study, in order to predict PR and AR of TBM, we considered these categories and selected the input parameters based on them. In the following sub-sections, the parameters related to these groups and their measurement process will be described.

3.4.1 Rock Material Properties

This sub-section describes the process to measure rock material properties to be used as input variables in LMR and NLMR modellings. In order to evaluate the rock material of the tunnel, extensive laboratory tests are required. For this purpose, more than 100 block samples were collected from the face of tunnel and after that were transferred to the laboratory for conducting relevant tests. It should be noted that these samples were collected from different tunnel distances (TDs) of three TBMs. In the following sections, procedures of the tests conducted in this research are presented.

Brazilian test is the most common indirect test to predict the tensile strength of the rock sample. Tensile strength indicates the strength of degree of bonding and interlocking between mineral grains that make up a rock sample. According to ISRM [54], Brazilian test specimen should be disc-shaped with an approximate ratio (diameter to thickness) of 2. The cylindrical surfaces should be free from any irregularities and end surfaces should be flat. In this study, various Brazilian tests were carried out on the disc-shaped samples of different TBMs. Based on ISRM [54], Brazilian tensile strength (BTS) values should be expressed in MPa. Values of 4.87 MPa and 15.68 MPa were obtained for the minimum and maximum BTS results, respectively in this study.

Rock compressive strength under uniaxial loading is assessed by the uniaxial compressive strength (UCS) test. Rock UCS can be obtained by loading the sample through load-controlled (e.g., 10 kN/s) or strain-controlled (e.g. 0.1% strain/s) procedures. Loading under strain-controlled processes is more dependent on loading conditions in the field and allows for obtaining a complete stress–strain curve of the sample. Many essential rock properties including Young's modulus can be specified based on the parameters monitored during the test. Specific samples' geometry was prepared as recommended by ISRM [54]. After coring and cutting the samples, their ends were made flat and perpendicular to the sample axis. The sides of samples were polished and smoothed, and the samples were checked to be without any crack, veins, fissures, and other flaws, which can influence the real properties of the rock. UCS tests were conducted on all collected samples. Minimum and maximum values of 72.3 MPa and 191.7 MPa were obtained for the UCS values in the laboratory.

3.4.2 Rock Mass Properties

In this study, a total TD of 5,443 m in different TBMs, which was divided into 560 panels (with approximately 10 m), was considered and observed in the site. For each panel, a mapping sheet with different information related to rock mass condition was prepared based on field observation. In each mapping sheet, several properties such as type of rock, rock mass strength, joint condition, and groundwater condition were observed. All mentioned properties were collected in various TDs of three TBMs. In each panel, rock quality designation (RQD) and rock mass rating (RMR) as two important rock mass properties were recorded. Ranges of 33.75–95% and 56.5–95 were achieved for RQD and RMR parameters, respectively, and were used in the modelling of TBM AR and TBM PR, in this study. It is important to note that all tunnel mapping and observations in this study were based on the suggested methods by ISRM [54].

3.4.3   Machine Characteristics

For the successful excavation of the hard rock, in this tunnelling project, the TBMs have been equipped using 35 numbers of cutters, which include 8 double cutters with a diameter of 17 inch and 27 single cutters with a diameter of 19 inch. By means of remote monitoring systems, the cutters are being monitored carefully for wear. The crew in wireless systems can arrange the cutter changes and keep track of wear through recording many variables upon each of the cutters. A sensor was provided for all cutters, which were bolted within the cutter housing. Using this mechanism, raw data can be transferred to a program that is displayed within the operator's cabin. In each panel, machine parameters such as PR, revolution per minute (RPM), AR, cutter head thrust force (TF), and cutter head torque were recorded by TBMs.

3.4.4   Input Variables

Results and reports of past relevant studies can help in selecting the most effective input parameters on TBM performance. As discussed before, UCS and BTS from the rock material group, RMR and RQD from the rock mass group and TF and RPM from machine characteristics were selected and used in this study. Each mentioned parameter represents a different issue and the provided information can help with a deeper understanding from the tunnelling project and operation [11]. Table 3.1

Table 3.1 The measured parameters in the site and laboratory together with their ranges and units

Parameter	Range	Mean	Unit
UCS	72.3–191.7	129.3	MPa
BTS	4.87–15.68	8.9	MPa
RMR	56.5–95	80.2	–
RQD	33.75–95	68.2	%
TF	91.3–565.8	263.2	KN
RPM	4.04–11.91	9.95	Rev/min
PR	0.12–5.8	2.5	m/h
AR	0.017–4	1	m/h

shows the measured parameters in the site and laboratory together with their ranges and units. In the following, simple and multiple regression analyses in the prediction of TBM AR and TBM PR are given.

3.5 Regression-Based Models

3.5.1 Simple Regression

As mentioned before, in this study, 560 data samples were prepared to propose simple regression, LMR and NLMR equations. In these data samples, RQD, UCS, RMR, BTS, TF, and RPM were set as input parameters to predict PR and AR of TBM. We examined various simple regression equations, i.e., power, linear, logarithmic, and exponential for each set of input–output. The proposed equations to estimate AR and PR of TBM are presented in Table 3.2. For evaluation of these equations, we used the coefficient of determination (R^2) in which its formula is described as follows:

$$R^2 = 1 - \left(\frac{\sum\limits_{i=1}^{n}(x_i - y_i)^2}{\sum\limits_{i=1}^{n}(x_i - \bar{x})^2} \right) \qquad (3.3)$$

where n means the total number of datasets, y_i means target value, and x_i and $\bar{x}$ are the predicted values and mean of the predicted values, respectively. The value for R^2 is considered as one for a perfect predictive model. In Table 3.2, the prediction

Table 3.2 Equations between inputs and PR/AR of TBM

Input	Output	Equation	R^2
RQD	PR	PR $= 4.426 \times \exp(-0.009 \times$ RQD$)$	0.145
UCS	PR	PR $= -0.011 \times$ UCS $+ 3.859$	0.221
RMR	PR	PR $= -0.026 \times$ RMR $+ 4.551$	0.118
BTS	PR	PR $= 3.324 \times \exp(-0.035 \times$ BTS$)$	0.142
TF	PR	PR $= 0.120 \times$ TF$^{0.538}$	0.187
RPM	PR	PR $= 0.489 \times$ RPM$^{0.708}$	0.232
RQD	AR	AR $= -0.021 \times$ RQD $+ 2.42$	0.234
UCS	AR	AR $= 4.207 \times \exp(-0.012 \times$ UCS$)$	0.351
RMR	AR	AR $= -0.025 \times$ RMR $+ 2.968$	0.114
BTS	AR	AR $= 2.076 \times \exp(-0.092 \times$ BTS$)$	0.212
TF	AR	AR $= 0.0001 \times$ TF$^{1.591}$	0.278
RPM	AR	AR $= 0.046 \times$ RPM$^{1.327}$	0.282

performances were evaluated using R^2. Generally, the results show low reliability of these input parameters to predict PR and AR of TBM. The maximum R^2 equal to 0.351 was obtained for the relationship between UCS and AR, which shows the importance of this parameter on TBM performance. The presented results in Table 3.2 reveal that new equations using more than a single input are needed to predict TBM performance with higher accuracy. Hence, in the following sub-section, the process of proposing LMR and NLMR models for TBM performance prediction will be presented.

3.5.2 Multiple Regression

LMR and NLMR analyses were conducted to predict PR and AR of TBM. To develop LMR and NLMR models, we decided to divide our database into five different train and test data samples. According to Swingler [55] and Looney [56], 20–25% of all the datasets could be used for testing while Nelson and Illingworth [57] stated that 20–30% of the whole datasets may be applied for testing. Hence, values of 448 and 112 data samples were selected as train and test datasets [58–64]. Using the constructed datasets, five LMR and five NLMR equations were proposed to predict TBM PR as listed in Table 3.3. As shown in this table, performance capacities (based on R^2) of NLMR models are higher than those of LMR models. This shows the capability of the NLMR models in predicting PR is higher compared to LMR models.

To select the best LMR and NLMR equations for PR prediction, Zorlu et al. [65] introduced a simple ranking method. The simple technique works by assigning a rank value to each performance index (Table 3.4). Each performance index was ordered in its class and the best performance index was assigned to the highest rating. For instance, R^2 values of 0.474, 0.498, 0.490, 0.493, and 0.480 were obtained for train 1 to train 5 of LMR technique, respectively. So, ratings of 1, 5, 3, 4, and 2 were assigned, respectively, and the same procedure was repeated for all performance indices and all predictive models (i.e., LMR and NLMR). The rating values of the LMR training dataset 1 were assigned as 1 for R^2, 4 for root mean square error (RMSE), and 1 for variance account for (VAF); thus, the performance rating was computed as 6 (Table 3.4). The final stage in the process of choosing the best model is to compute the total rank by summing up the rank value of each dataset (training and testing) according to the list in Table 3.5. Based on this table, models No. 4 (with a total rank of 23) and 2 (with a total rank of 29) exhibited the best performance capacities for LMR and NLMR models, respectively. The equations for the calculation of RMSE and VAF are presented as follows:

Table 3.3 Proposed LMR and NLMR equations in predicting TBM PR

Model	Dataset No.	Proposed equation	R^2
LMR	1	$PR = -0.012 \times RQD - 0.007 \times UCS - 0.015 \times RMR - 0.029 \times BTS + 0.003 \times TF + 0.059 \times RPM + 4.387$	0.474
	2	$PR = -0.014 \times RQD - 0.007 \times UCS - 0.013 \times RMR - 0.022 \times BTS + 0.003 \times TF + 0.052 \times RPM + 4.372$	0.498
	3	$PR = -0.016 \times RQD - 0.007 \times UCS - 0.016 \times RMR - 0.022 \times BTS + 0.003 \times TF + 0.049 \times RPM + 4.754$	0.490
	4	$PR = -0.013 \times RQD - 0.007 \times UCS - 0.014 \times RMR - 0.021 \times BTS + 0.003 \times TF + 0.057 \times RPM + 4.291$	0.493
	5	$PR = -0.012 \times RQD - 0.006 \times UCS - 0.013 \times RMR - 0.028 \times BTS + 0.003 \times TF + 0.045 \times RPM + 4.122$	0.480
NLMR	1	$PR = 2.394 \times \exp(-0.008 \times RQD) + 800.455 \times UCS^{-1.5} - 0.015 \times RMR + 24.992 \times \exp(-0.8 \times BTS) + 0.003 \times TF^{0.932} + 1.583 \times RPM^{0.219} - 1.672$	0.538
	2	$PR = 13.906 \times \exp(-0.001 \times RQD) + 817.189 \times UCS^{-1.5} - 0.013 \times RMR + 0.745 \times \exp(-0.039 \times BTS) + 0.003 \times TF^{0.984} + 0.631 \times RPM^{0.343} - 12.631$	0.545
	3	$PR = 3.035 \times \exp(-0.009 \times RQD) + 241.291 \times UCS^{-1.2} - 0.018 \times RMR + 1.364 \times BTS^{-0.31} + 0.003 \times TF + 0.138 \times RPM^{0.665} - 0.605$	0.538
	4	$PR = 2.392 \times \exp(-0.009 \times RQD) + 543.883 \times UCS^{-1.4} - 0.015 \times RMR + 0.760 \times \exp(-0.037 \times BTS) + 0.001 \times TF^{1.2} + 0.133 \times RPM^{0.741} - 0.168$	0.534
	5	$PR = 12.567 \times \exp(-0.001 \times RQD) + 722.487 \times UCS^{-1.5} - 0.014 \times RMR + 0.958 \times \exp(-0.038 \times BTS) + 0.002 \times TF^{1.03} + 0.109 \times RPM^{0.743} - 10.733$	0.522

$$VAF = \left[1 - \frac{\text{var}(y - y\prime)}{\text{var}(y)} \right] \times 100 \tag{3.4}$$

$$RMSE = \sqrt{\frac{1}{N} \sum_{i=1}^{N} (y - y\prime)^2} \tag{3.5}$$

Table 3.4 Performance indices of LMR and NLMR models and their rank values in predicting TBM PR

Method	Model	R^2	RMSE	VAF (%)	Rating for R^2	Rating for RMSE	Rating for VAF	Rank value
LMR	Train 1	0.474	0.533	47.538	1	4	1	6
	Train 2	0.498	0.529	47.775	5	5	2	12
	Train 3	0.490	0.534	49.034	3	3	4	10
	Train 4	0.493	0.529	49.303	4	5	5	14
	Train 5	0.480	0.533	47.956	2	4	3	9
	Test 1	0.498	0.464	47.627	4	5	5	14
	Test 2	0.340	0.480	31.731	1	3	1	5
	Test 3	0.520	0.487	37.587	5	2	2	9
	Test 4	0.389	0.478	37.948	2	4	3	9
	Test 5	0.455	0.464	45.355	3	5	4	12
NLMR	Train 1	0.538	0.501	53.786	4	5	4	13
	Train 2	0.545	0.504	54.471	5	4	5	14
	Train 3	0.538	0.508	53.750	4	3	3	10
	Train 4	0.534	0.576	52.870	3	1	2	6
	Train 5	0.522	0.520	52.136	2	2	1	5
	Test 1	0.512	0.499	42.988	2	2	2	6
	Test 2	0.522	0.405	51.790	5	5	5	15
	Test 3	0.520	0.487	37.587	4	3	1	8
	Test 4	0.505	0.559	48.289	1	1	3	5
	Test 5	0.517	0.446	51.515	3	4	4	11

Table 3.5 Total rank values for LMR and NLMR techniques in predicting TBM PR

Method	Model	Total rank
LMR	1	20
	2	17
	3	19
	4	23
	5	21
NLMR	1	19
	2	29
	3	18
	4	11
	5	16

in these equations, y and y' are the predicted and measured values, respectively, and the total number of data showed by N. VAF $=100\%$ and RMSE $= 0$ will be the ideal condition for a predictive model. Equations of models No. 4 and 2 in Table 3.5 for LMR and NLMR techniques, respectively, are as follows:

$$PR = -0.013 \times RQD - 0.007 \times UCS - 0.014 \times RMR - 0.021 \times BTS + 0.003 \times TF + 0.057 \times RPM + 4.291$$

(3.6)

$$PR = 13.906 \times \exp(-0.001 \times RQD) + 817.189 \times UCS^{-1.5} - 0.013 \times RMR + 0.745 \times \exp(-0.039 \times BTS) + 0.003 \times TF^{0.984} + 0.631 \times RPM^{0.343} - 12.631$$

(3.7)

The graphs of predicted PR using LMR and NLMR techniques against the measured PR for training and testing datasets are shown in Figs. 3.2 and 3.3, respectively. Based on these figures, NLMR model shows higher capability for predicting PR compared to LMR model.

Considering the explained procedure for PR prediction and using five randomly selected datasets, five LMR and five NLMR equations were developed to predict TBM AR as indicated in Table 3.6. Based on this table, in overall, R^2 values of NLMR equations are better than those of LMR models. A ranking system was also used and calculated here for each training and testing dataset (Table 3.7). After that, assigned ratings of R^2, RMSE, and VAF were summed up for each training and testing datasets (Table 3.8). According to total rank values, models No. 3 in both LMR and NLMR techniques showed the best prediction capabilities. Relevant equations of these models are presented as follows:

$$AR = -0.012 \times RQD - 0.005 \times UCS - 0.004 \times RMR - 0.021 \times BTS + 0.003 \times TF + 0.049 \times RPM + 1.808$$

(3.8)

$$AR = -0.009 \times RQD - 0.005 \times UCS - 0.001 \times RMR + 0.716 \times \exp(-0.053 \times BTS) + 0.00001 \times TF^{2.167} + 0.026 \times RPM^{1.234} + 1.179$$

(3.9)

Figures 3.4 and 3.5 illustrate the measured AR values against predicted AR values using LMR and NLMR techniques for training and testing datasets, respectively. In these figures, the R^2 equal to 0.661 for testing dataset of NLMR model suggests the superiority of this technique in predicting AR, while this value obtained for the LMR model is 0.582.

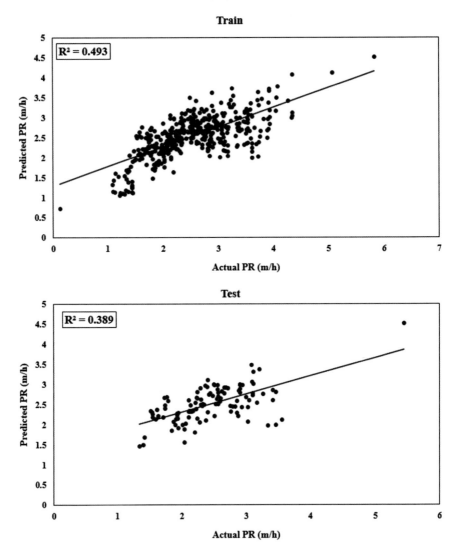

Fig. 3.2 Predicted PR using LMR technique for training and testing datasets

3.6 Discussion and Conclusion

Simple, LMR, and NLMR techniques were used and developed to produce statistical equations for predicting TBM performance. A range of 0.11–0.35 was obtained for R^2 of simple regression equations where AR/PR was our model output. Meaning to say that the highest R^2 value using a single input is equal to 0.35, which is considered as a low capability of these techniques. Due to that and in order to get a higher level of accuracy, we decided to use six model inputs, i.e., RQD, UCS, RMR, BTS, TF, and

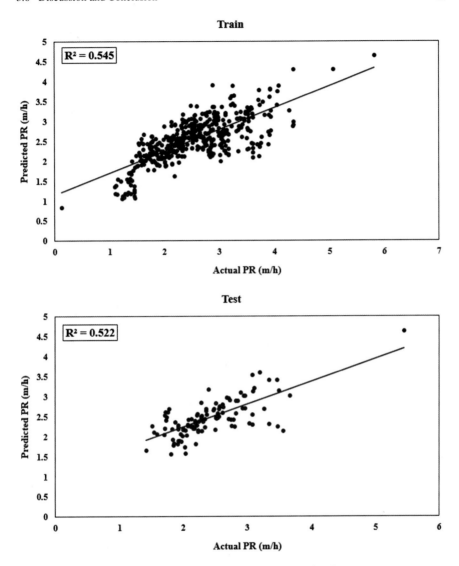

Fig. 3.3 Predicted PR using NLMR technique for training and testing datasets

RPM instead of single model input. Then, five LMR and five NLMR equations were developed and their performance capacities were assessed using three performance indices, i.e., R^2, VAF, and RMSE. In addition, since selecting the best LMR and NLMR equations was difficult due to their close performance indices, we applied a simple ranking procedure suggested by Zorlu et al. [65]. Generally, it was found that the proposed models are able to predict TBM performance parameters with suitable degree of accuracy. Several studies mentioned that R^2 values ranging from 0.5 to 0.7

Table 3.6 Proposed LMR and NLMR equations in predicting TBM AR

Model	Dataset No.	Proposed equation	R^2
LMR	1	$AR = -0.014 \times RQD - 0.005 \times UCS - 0.003 \times RMR - 0.016 \times BTS + 0.003 \times TF + 0.054 \times RPM + 1.785$	0.586
	2	$AR = -0.012 \times RQD - 0.004 \times UCS - 0.010 \times RMR - 0.021 \times BTS + 0.003 \times TF + 0.058 \times RPM + 2.196$	0.583
	3	$AR = -0.012 \times RQD - 0.005 \times UCS - 0.004 \times RMR - 0.021 \times BTS + 0.003 \times TF + 0.049 \times RPM + 1.808$	0.584
	4	$AR = -0.011 \times RQD - 0.005 \times UCS - 0.008 \times RMR - 0.016 \times BTS + 0.003 \times TF + 0.054 \times RPM + 1.804$	0.580
	5	$AR = -0.012 \times RQD - 0.005 \times UCS - 0.007 \times RMR - 0.023 \times BTS + 0.003 \times TF + 0.055 \times RPM + 1.928$	0.569
NLMR	1	$AR = -0.011 \times RQD + 1.862 \times \exp(-0.008 \times UCS) - 0.003 \times RMR + 1.063 \times \exp(-0.018 \times BTS) + 0.00001 \times TF^{1.887} + 0.024 \times RPM^{1.282} - 0.392$	0.600
	2	$AR = -0.013 \times RQD + 1.596 \times \exp(-0.009 \times UCS) - 0.008 \times RMR + 1.361 \times \exp(-0.018 \times BTS) + 0.00002 \times TF^{1.797} + 0.022 \times RPM^{1.335} + 0.018$	0.594
	3	$AR = -0.009 \times RQD - 0.005 \times UCS - 0.001 \times RMR + 0.716 \times \exp(-0.053 \times BTS) + 0.00001 \times TF^{2.167} + 0.026 \times RPM^{1.234} + 1.179$	0.604
	4	$AR = -0.009 \times RQD + 1.781 \times \exp(-0.008 \times UCS) - 0.007 \times RMR + 0.912 \times \exp(-0.022 \times BTS) + 0.00001 \times TF^{1.908} + 0.024 \times RPM^{1.285} + 0.263$	0.600
	5	$AR = -0.009 \times RQD + 1.754 \times \exp(-0.008 \times UCS) - 0.007 \times RMR + 1.814 \times \exp(-0.015 \times BTS) + 0.00001 \times TF^{1.856} + 0.024 \times RPM^{1.294} + 0.263$	0.583

are suitable acceptance in the field of TBM performance prediction [38, 66]. Nevertheless, NLMR equations show higher capability in predicting both TBM PR/AR values. Specifically, considering testing data samples, values of 0.389, 0.478, and 37.948% and 0.522, 0.405, and 51.790% were obtained for R^2, RMSE, and VAF of the best LMR and NLMR equations in predicting TBM PR. Similarly, these values were obtained as 0.582, 0.360, and 58.180% and 0.661, 0.328, and 65.659% in estimating TBM AR. These results revealed that the proposed AR equations work better

Table 3.7 Performance indices of LMR and NLMR models and their rank values in predicting TBM AR

Method	Model	R^2	RMSE	VAF (%)	Rating for R^2	Rating for RMSE	Rating for VAF	Rank value
LMR	Train 1	0.586	0.373	58.633	5	4	5	14
	Train 2	0.583	0.366	58.333	3	5	3	11
	Train 3	0.584	0.379	58.423	4	2	4	10
	Train 4	0.580	0.378	57.976	2	3	2	7
	Train 5	0.569	0.378	56.913	1	3	1	5
	Test 1	0.568	0.392	56.599	1	3	1	5
	Test 2	0.581	0.415	58.079	2	2	2	6
	Test 3	0.582	0.360	58.180	3	5	3	11
	Test 4	0.620	0.368	61.290	4	3	4	11
	Test 5	0.629	0.361	62.866	5	4	5	14
NLMR	Train 1	0.600	0.368	60.015	4	4	4	12
	Train 2	0.594	0.361	59.424	3	5	2	10
	Train 3	0.604	0.371	60.302	5	2	5	12
	Train 4	0.600	0.369	59.979	4	3	3	10
	Train 5	0.583	0.372	58.270	2	1	1	4
	Test 1	0.607	0.378	60.702	1	2	1	4
	Test 2	0.612	0.400	61.076	2	1	2	5
	Test 3	0.661	0.328	65.659	5	5	5	15
	Test 4	0.624	0.370	61.279	3	3	3	9
	Test 5	0.651	0.352	64.835	4	4	4	12

Table 3.8 Total rank for LMR and NLMR techniques in predicting TBM AR

Method	Model	Total rank
LMR	1	19
	2	17
	3	21
	4	18
	5	19
NLMR	1	16
	2	15
	3	27
	4	19
	5	16

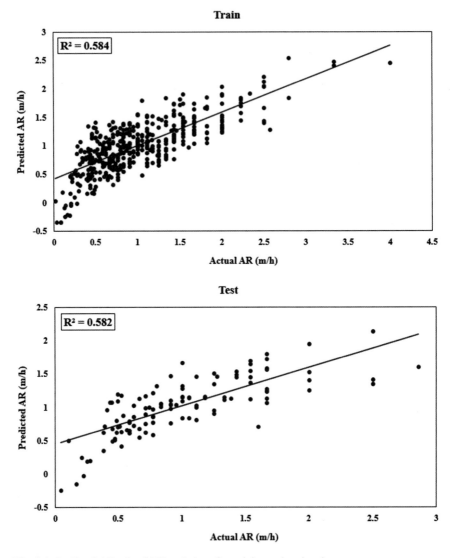

Fig. 3.4 Predicted AR using LMR technique for training and testing datasets

than the proposed PR equations. As expected, NLMR models are able to have a better mapping between input and output parameters. It is worth noting that, in practice, all proposed methods have the capability of the PR and AR predictions. However, in order to minimize the obtained system error, new computational techniques such as artificial intelligence and machine learning can be very useful and more applicable in the area of tunnelling and TBM performance.

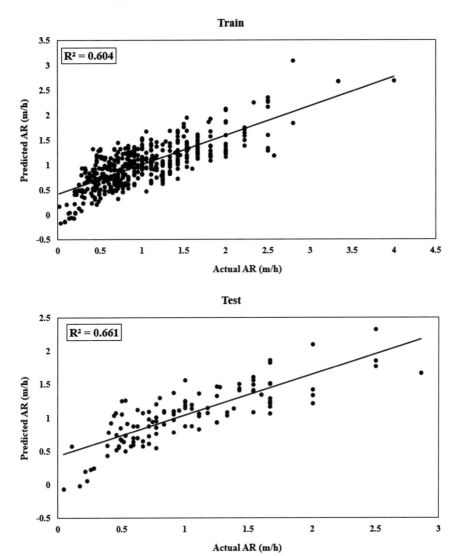

Fig. 3.5 Predicted AR using NLMR technique for training and testing datasets

References

1. S. Yagiz, Development of rock fracture and brittleness indices to quantify the effects of rock mass features and toughness in the CSM Model basic penetration for hard rock tunneling machines (2002)
2. S. Yagiz, Utilizing rock mass properties for predicting TBM performance in hard rock condition. Tunn. Undergr. Sp. Technol. **23**, 326–339 (2008)
3. Z.T. Bieniawski, B. Celada, J.M. Galera, M. Álvares, Rock mass excavability (RME) index, in *ITA World Tunnel Congress* (Korea, 2006)

4. J. Zeng, B. Roy, D. Kumar, A.S. Mohammed, D.J. Armaghani, J. Zhou, E.T. Mohamad, Proposing several hybrid PSO-extreme learning machine techniques to predict TBM performance. Eng. Comput. (n.d.). https://doi.org/10.1007/s00366-020-01225-2

5. D.J. Armaghani, E.T. Mohamad, M.S. Narayanasamy, N. Narita, S. Yagiz, Development of hybrid intelligent models for predicting TBM penetration rate in hard rock condition. Tunn. Undergr. Sp. Technol. **63**, 29–43 (2017). https://doi.org/10.1016/j.tust.2016.12.009

6. J. Zhou, B. Yazdani Bejarbaneh, D. Jahed Armaghani, M.M. Tahir, Forecasting of TBM advance rate in hard rock condition based on artificial neural network and genetic programming techniques. Bull. Eng. Geol. Environ. **79**, 2069–2084 (2020). https://doi.org/10.1007/s10064-019-01626-8

7. J. Zhou, Y. Qiu, D.J. Armaghani, W. Zhang, C. Li, S. Zhu, R. Tarinejad, Predicting TBM penetration rate in hard rock condition: a comparative study among six XGB-based metaheuristic techniques. Geosci. Front. (2020). https://doi.org/10.1016/j.gsf.2020.09.020

8. J. Zhou, Y. Qiu, S. Zhu, D.J. Armaghani, M. Khandelwal, E.T. Mohamad, Estimation of the TBM advance rate under hard rock conditions using XGBoost and Bayesian optimization. Undergr. Sp. (2020). https://doi.org/10.1016/j.undsp.2020.05.008

9. M.A. Grima, P.A. Bruines, P.N.W. Verhoef, Modeling tunnel boring machine performance by neuro-fuzzy methods. Tunn. Undergr. Sp. Technol. **15**, 259–269 (2000)

10. S. Yagiz, C. Gokceoglu, E. Sezer, S. Iplikci, Application of two non-linear prediction tools to the estimation of tunnel boring machine performance. Eng. Appl. Artif. Intell. **22**, 808–814 (2009)

11. J.K. Hamidi, K. Shahriar, B. Rezai, J. Rostami, Performance prediction of hard rock TBM using Rock Mass Rating (RMR) system. Tunn. Undergr. Sp. Technol. **25**, 333–345 (2010)

12. S. Mahdevari, K. Shahriar, S. Yagiz, M.A. Shirazi, A support vector regression model for predicting tunnel boring machine penetration rates. Int. J. Rock Mech. Min. Sci. **72**, 214–229 (2014)

13. R.K. Goel, Evaluation of TBM performance in a Himalayan tunnel, in: Proc. World Tunn. Congr. India, 2008: pp. 1522–1532

14. A. Palmstrom, E. Broch, Use and misuse of rock mass classification systems with particular reference to the Q-system. Tunn. Undergr. Sp. Technol. **21**, 575–593 (2006)

15. N. Innaurato, A. Mancini, E. Rondena, A. Zaninetti, Forecasting and effective TBM performances in a rapid excavation of a tunnel in Italy, in *7th ISRM Congress, International Society for Rock Mechanics and Rock Engineering* (1991)

16. S.V.A.N.K. Abad, A. Tugrul, C. Gokceoglu, D.J. Armaghani, Characteristics of weathering zones of granitic rocks in Malaysia for geotechnical engineering design. Eng. Geol. **200**, 94–103 (2016)

17. E. Tonnizam Mohamad, D. Jahed Armaghani, M. Ghoroqi, B. Yazdani Bejarbaneh, T. Ghahremanians, M.Z. Abd Majid, O. Tabrizi, Ripping production prediction in different weathering zones according to field data. Geotech. Geol. Eng. **35** (2017). https://doi.org/10.1007/s10706-017-0254-4

18. E.T. Mohamad, D. Li, B.R. Murlidhar, D.J. Armaghani, K.A. Kassim, I. Komoo, The effects of ABC, ICA, and PSO optimization techniques on prediction of ripping production. Eng. Comput. (2019). https://doi.org/10.1007/s00366-019-00770-9

19. D. Jahed Armaghani, M.F. Mohd Amin, S. Yagiz, R.S. Faradonbeh, R.A. Abdullah, Prediction of the uniaxial compressive strength of sandstone using various modeling techniques. Int. J. Rock Mech. Min. Sci. **85**, 174–186 (2016). https://doi.org/10.1016/j.ijrmms.2016.03.018

20. E. Momeni, R. Nazir, D.J. Armaghani, E.T. Mohamad, Prediction of unconfined compressive strength of rocks: a review paper. J. Teknol. **77** (2015)

21. B.Y. Bejarbaneh, E.Y. Bejarbaneh, A. Fahimifar, D.J. Armaghani, M.Z.A. Majid, Intelligent modelling of sandstone deformation behaviour using fuzzy logic and neural network systems. Bull. Eng. Geol. Environ. **77**, 345–361 (2018)

22. B. Gordan, D.J. Armaghani, A.B. Adnan, A.S.A. Rashid, A new model for determining slope stability based on seismic motion performance. Soil Mech. Found. Eng. **53**, 344–351 (2016). https://doi.org/10.1007/s11204-016-9409-1

23. R.S. Faradonbeh, D.J. Armaghani, M. Monjezi, Development of a new model for predicting flyrock distance in quarry blasting: a genetic programming technique. Bull. Eng. Geol. Environ. **75**, 993–1006 (2016)

24. M. Hasanipanah, R.S. Faradonbeh, H.B. Amnieh, D.J. Armaghani, M. Monjezi, Forecasting blast-induced ground vibration developing a CART model. Eng. Comput. 1–10 (2016)

25. D.J. Armaghani, M. Hasanipanah, H.B. Amnieh, E.T. Mohamad, Feasibility of ICA in approximating ground vibration resulting from mine blasting. Neural Comput. Appl. **29**, 457–465 (2018)

26. M. Hasanipanah, D.J. Armaghani, H.B. Amnieh, M.Z.A. Majid, M.M.D. Tahir, Application of PSO to develop a powerful equation for prediction of flyrock due to blasting. Neural Comput. Appl. **28**, 1043–1050 (2017)

27. M. Khandelwal, D.J. Armaghani, R.S. Faradonbeh, P.G. Ranjith, S. Ghoraba, A new model based on gene expression programming to estimate air flow in a single rock joint. Environ. Earth Sci. **75**, 739 (2016)

28. D.J. Armaghani, E.T. Mohamad, M. Hajihassani, S. Yagiz, H. Motaghedi, Application of several non-linear prediction tools for estimating uniaxial compressive strength of granitic rocks and comparison of their performances. Eng. Comput. **32**, 189–206 (2016)

29. R. Shirani Faradonbeh, D. Jahed Armaghani, M.Z. Abd Majid, M. MD Tahir, B. Ramesh Murlidhar, M. Monjezi, H.M. Wong, Prediction of ground vibration due to quarry blasting based on gene expression programming: a new model for peak particle velocity prediction. Int. J. Environ. Sci. Technol. **13** (2016). https://doi.org/10.1007/s13762-016-0979-2

30. D.J. Armaghani, A. Mahdiyar, M. Hasanipanah, R.S. Faradonbeh, M. Khandelwal, H.B. Amnieh, Risk assessment and prediction of flyrock distance by combined multiple regression analysis and monte carlo simulation of quarry blasting. Rock Mech. Rock Eng. **49**, 1–11 (2016). https://doi.org/10.1007/s00603-016-1015-z

31. A. Mahdiyar, M. Hasanipanah, D.J. Armaghani, B. Gordan, A. Abdullah, H. Arab, M.Z.A. Majid, A Monte Carlo technique in safety assessment of slope under seismic condition. Eng. Comput. **33**, 807–817 (2017). https://doi.org/10.1007/s00366-016-0499-1

32. D.J. Armaghani, R.S. Faradonbeh, H. Rezaei, A.S.A. Rashid, H.B. Amnieh, Settlement prediction of the rock-socketed piles through a new technique based on gene expression programming. Neural Comput. Appl. **29**, 1115–1125 (2016). https://doi.org/10.1007/s00521-016-2618-8

33. M. Liang, E.T. Mohamad, R.S. Faradonbeh, D. Jahed Armaghani, S. Ghoraba, Rock strength assessment based on regression tree technique. Eng. Comput. **32**, 343–354 (2016). https://doi.org/10.1007/s00366-015-0429-7

34. M. Monjezi, M. Baghestani, R. Shirani Faradonbeh, M. Pourghasemi Saghand, D. Jahed Armaghani, Modification and prediction of blast-induced ground vibrations based on both empirical and computational techniques. Eng. Comput. **32** (2016). https://doi.org/10.1007/s00366-016-0448-z

35. M. Khandelwal, R.S. Faradonbeh, M. Monjezi, D.J. Armaghani, M.Z.B.A. Majid, S. Yagiz, Function development for appraising brittleness of intact rocks using genetic programming and non-linear multiple regression models. Eng. Comput. **33**, 13–21 (2017)

36. M. Khandelwal, M. Monjezi, Prediction of backbreak in open-pit blasting operations using the machine learning method. Rock Mech. Rock Eng. **46**, 389–396 (2013)

37. Q.-M. Gong, J. Zhao, Development of a rock mass characteristics model for TBM penetration rate prediction. Int. J. Rock Mech. Min. Sci. **46**, 8–18 (2009)

38. E. Farrokh, J. Rostami, C. Laughton, Study of various models for estimation of penetration rate of hard rock TBMs. Tunn. Undergr. Sp. Technol. **30**, 110–123 (2012)

39. M. Koopialipoor, S.S. Nikouei, A. Marto, A. Fahimifar, D.J. Armaghani, E.T. Mohamad, Predicting tunnel boring machine performance through a new model based on the group method of data handling. Bull. Eng. Geol. Environ. **78**, 3799–3813 (2018)

40. D.J. Armaghani, V. Safari, A. Fahimifar, M. Monjezi, M.A. Mohammadi, Uniaxial compressive strength prediction through a new technique based on gene expression programming. Neural Comput. Appl. **30**, 3523–3532 (2018)

41. D.J. Armaghani, E.T. Mohamad, E. Momeni, M. Monjezi, M.S. Narayanasamy, Prediction of the strength and elasticity modulus of granite through an expert artificial neural network. Arab. J. Geosci. **9**, 48 (2016)
42. M. Hasanipanah, R.S. Faradonbeh, D.J. Armaghani, H.B. Amnieh, M. Khandelwal, Development of a precise model for prediction of blast-induced flyrock using regression tree technique. Environ. Earth Sci. **76**, 27 (2017)
43. M.A. Grima, R. Babuška, Fuzzy model for the prediction of unconfined compressive strength of rock samples. Int. J. Rock Mech. Min. Sci. **36**, 339–349 (1999)
44. C. Gokceoglu, K. Zorlu, A fuzzy model to predict the uniaxial compressive strength and the modulus of elasticity of a problematic rock. Eng. Appl. Artif. Intell. **17**, 61–72 (2004)
45. D.J. Armaghani, E.T. Mohamad, E. Momeni, M.S. Narayanasamy, An adaptive neuro-fuzzy inference system for predicting unconfined compressive strength and Young's modulus: a study on Main Range granite. Bull. Eng. Geol. Environ. **74**, 1301–1319 (2015)
46. A. Bahrami, M. Monjezi, K. Goshtasbi, A. Ghazvinian, Prediction of rock fragmentation due to blasting using artificial neural network. Eng. Comput. **27**, 177–181 (2011)
47. D. Jahed Armaghani, M. Hajihassani, B. Yazdani Bejarbaneh, A. Marto, E. Tonnizam Mohamad, Indirect measure of shale shear strength parameters by means of rock index tests through an optimized artificial neural network. Meas. J. Int. Meas. Confed. **55**, 487–498 (2014). https://doi.org/10.1016/j.measurement.2014.06.001
48. M. Hajihassani, D. Jahed Armaghani, A. Marto, E. Tonnizam Mohamad, Ground vibration prediction in quarry blasting through an artificial neural network optimized by imperialist competitive algorithm. Bull. Eng. Geol. Environ. **74**, 873–886 (2014). https://doi.org/10.1007/s10064-014-0657-x
49. M. Hasanipanah, D. Jahed Armaghani, M. Monjezi, S. Shams, Risk assessment and prediction of rock fragmentation produced by blasting operation: a rock engineering system. Environ. Earth Sci. **75** (2016). https://doi.org/10.1007/s12665-016-5503-y
50. S. Yagiz, C. Gokceoglu, Application of fuzzy inference system and nonlinear regression models for predicting rock brittleness. Expert Syst. Appl. **37**, 2265–2272 (2010)
51. E.T. Mohamad, D.J. Armaghani, A. Mahdyar, I. Komoo, K.A. Kassim, A. Abdullah, M.Z.A. Majid, Utilizing regression models to find functions for determining ripping production based on laboratory tests. Measurement **111**, 216–225 (2017)
52. Z. Nordin, Planning and construction of pahang-selangor raw water transfer (PSRWT) Tunnel, in Seminar Tunnels Undergrated Structure 2–4 Sept. (Kuala Lumpur, Malaysia, 2014)
53. D.J. Armaghani, M. Koopialipoor, A. Marto, S. Yagiz, Application of several optimization techniques for estimating TBM advance rate in granitic rocks. J. Rock Mech. Geotech. Eng. (2019). https://doi.org/10.1016/j.jrmge.2019.01.002
54. R. Ulusay, J.A. Hudson, ISRM The complete ISRM suggested methods for rock characterization, testing and monitoring: 1974–2006. Comm. Test. Methods Int. Soc. Rock Mech. Compil. **628**(n.d.). (Arranged by ISRM Turkish Natl. Group, Ankara, Turkey, 2007)
55. K. Swingler, *Applying Neural Networks: A Practical Guide* (Academic Press, New York, 1996)
56. C.G. Looney, Advances in feedforward neural networks: demystifying knowledge acquiring black boxes. IEEE Trans. Knowl. Data Eng. **8**, 211–226 (1996)
57. M.M. Nelson, W.T. Illingworth, *A Practical Guide to Neural Nets* (Addison-Wesley Reading, MA, 1991)
58. D.J. Armaghani, M. Hajihassani, E.T. Mohamad, A. Marto, S.A. Noorani, Blasting-induced flyrock and ground vibration prediction through an expert artificial neural network based on particle swarm optimization. Arab. J. Geosci. **7**, 5383–5396 (2014)
59. D. Li, D.J. Armaghani, J. Zhou, S.H. Lai, M. Hasanipanah, A GMDH predictive model to predict rock material strength using three non-destructive tests. J. Nondestruct. Eval. (2020). https://doi.org/10.1007/s10921-020-00725-x
60. H. Harandizadeh, D.J. Armaghani, Prediction of air-overpressure induced by blasting using an ANFIS-PNN model optimized by GA. Appl. Soft Comput. 106904 (2020). https://doi.org/10.1016/j.asoc.2020.106904

61. D.J. Armaghani, E. Momeni, P.G. Asteris, Application of group method of data handling technique in assessing deformation of rock mass. Metaheuristic Comput. Appl. **1**, 1–18 (2020)
62. E. Momeni, M.B. Dowlatshahi, F. Omidinasab, H. Maizir, D.J. Armaghani, Gaussian process regression technique to estimate the pile bearing capacity. Arab. J. Sci. Eng. **45**, 8255–8267 (2020). https://doi.org/10.1007/s13369-020-04683-4
63. E. Momeni, A. Yarivand, M.B. Dowlatshahi, D.J. Armaghani, An efficient optimal neural network based on gravitational search algorithm in predicting the deformation of geogrid-reinforced soil structures. Transp. Geotech. 100446 (2020)
64. J. Huang, M. Koopialipoor, D.J. Armaghani, A combination of fuzzy Delphi method and hybrid ANN-based systems to forecast ground vibration resulting from blasting. Sci. Rep. **10**, 1–21 (2020)
65. K. Zorlu, C. Gokceoglu, F. Ocakoglu, H.A. Nefeslioglu, S. Acikalin, Prediction of uniaxial compressive strength of sandstones using petrography-based models. Eng. Geol. **96**, 141–158 (2008)
66. A. Salimi, M. Esmaeili, Utilising of linear and non-linear prediction tools for evaluation of penetration rate of tunnel boring machine in hard rock condition. Int. J. Min. Miner. Eng. **4**, 249–264 (2013)

Chapter 4
A Comparative Study of Artificial Intelligence Techniques to Estimate TBM Performance in Various Weathering Zones

Abstract This study aims to propose a practical intelligence way for the prediction of tunnel boring machine (TBM) performance in various weathering zones. To do this, after reviewing the available literature, the data collected from the tunnel site and doing laboratory investigations, five important parameters, i.e., rock mass rating, Brazilian tensile strength, weathering zone, cutter head thrust force, and revolution per minute, were set as model inputs to predict penetration rate (PR) of TBM. Then, two intelligence techniques, namely, group method of data handling (GMDH) and artificial neural network (ANN) were applied to the collected data (i.e., 202 data samples). In developing these intelligence techniques, a series of parametric studies were conducted on the most important parameters of these techniques. After developing GMDH and ANN models, some important performance indices were selected and calculated to select the best one among them. It was found that the GMDH model receives a higher accuracy level compared to the ANN model. It can be established that the GMDH is an applicable and powerful technique in the area of TBM and tunnelling technology.

Keywords Group method of data handling · Tunnel boring machine · Tunnelling technology · Artificial neural network · Predictive model

4.1 Introduction

One of the most significant objectives of tunnel engineering is enhancing the performance quality of drilling and boring systems in various situations and working conditions. A major issue that engineers take into consideration is the tunnel boring machine (TBM) performance; this is useful in testing the parameters influencing the performance of a drilling system. Thus, we can significantly reduce the risks accompanying tunnelling projects by appropriately estimating the excavating system [1–3]. The discussion that arose in relation to the effective features of TBMs in relevant literature led to a number of studies conducted on this issue, resulting in the proposal of a variety of empirical and field models in this field [4–8]. These models consider a limited number of parameters to have an impact on the TBMs performance; at the

© The Author(s), under exclusive license to Springer Nature Singapore Pte Ltd. 2021
D. Jahed Armaghani and A. Azizi, *Applications of Artificial Intelligence in Tunnelling and Underground Space Technology*, SpringerBriefs in Applied Sciences and Technology, https://doi.org/10.1007/978-981-16-1034-9_4

same time, they do not consider many significant parameters [9–12]. As a result, the models introduced in the literature generally lack the precision that is needed for TBMs to perform at a high level [13–16].

On the other hand, the flexible nature of machine learning (ML), artificial intelligence (AI), and soft computing techniques make these techniques as powerful, applicable, and easy to conduct in solving engineering and science problems [17–35]. Many AI models have been suggested for the prediction of TBM performance [1, 13, 16, 36–45]. Some of these models include particle swarm optimization (PSO), fuzzy logic, artificial neural network (ANN), adoptive neuro-fuzzy inference system (ANFIS), imperialist competitive algorithm (ICA), etc. They have been applied to approximate TBM performance parameters such as advance rate (AR) and penetration rate (PR). An ANFIS model was proposed by Grima et al. [46] for predicting PR and it was confirmed that ANFIS rendered a more accurate performance in comparison with the existing statistical methods. Mahdevari et al. [47] proposed a support vector regression (SVR) model, which was a machine learning one. A hybrid ANFIS–fuzzy C-means clustering approach was developed and used by Fattahi [48] for the assessment of TBM PR. Minh et al. [49] introduced a new fuzzy model to predict TBM PR. Zhou et al. [16, 41] suggested two applicable AI techniques on the basis of support vector machine, and genetic programming for forecasting TBM AR. A review of the mentioned studies shows that AI techniques are considered as feasible and capable methods in the area of tunnelling.

This study aims to conduct and develop an AI technique, namely, the group method of data handling (GMDH) to predict TBM PR in various weathering zones. To show the capability of the GMDH model in prediction TBM performance, an ANN model is also built and proposed. The results obtained by these two models will be assessed and compared to select and introduce the best AI model in the area of TBM and tunnelling.

4.2 Methodology

4.2.1 Artificial Neural Network (ANN)

Artificial neural network (ANN), nowadays, is a well-known method for analyzing complicated engineering problems. In this method, a mathematical model of reasoning is defined by simulating the biological nervous systems [40, 50, 51]. ANN would develop patterns using actual data and would able to process the information so that they can estimate a relationship among variables [52]. They learn the connection between parameters from examples and can expand it for new inputs with an acceptable accuracy.

ANNs have two types of structure means: recurrent and feed-forward. In feed-forward ANNs, like multi-layer perceptron (MLP) ANNs, there are several layers of neurons, which lead a signal from input layer(s) to the outputs by connections

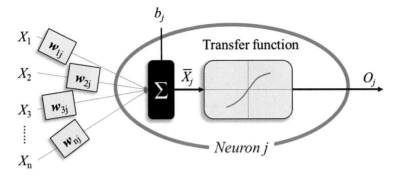

Fig. 4.1 A view of neuron j in an ANN model

[53–55]. In recurrent ANNs, signals can move both forward to neurons in past layers or backward to feedback to the same neuron. However, the most significant property of ANNs is utilizing a learning algorithm to learn from samples and improve their performance similar to the human brain. The most common training algorithm is the backpropagation algorithm. It reduces the error between input and output data during the learning process until it can adjust the network weights [56–59]. Using both forward and backward flows and error correction in various layers, BP learning tries to reach the outputs of the network to the desired values. Figure 4.1 illustrates a mathematical model of an artificial neuron.

4.2.2 Group Method of Data Handling (GMDH)

The GMDH method is from a type of algorithms, which are well known for being self-organized. GMDH benefits from the input and output datasets to model the behaviour of a phenomenon. Similar to most of the artificial neural networks, dataset for GMDH comprises a pair of data composed of several inputs and one corresponding output. In fact, every neuron layer in GMDH model has several data pairs that are linked to each other and eventually link Input data to output.

Ivahnenko [60] was the pioneer who introduced the first self-organizing GMDH algorithm, which has the ability to generate quadratic polynomials in any neuron. Therefore, it can filter partial descriptions (PD's) with the best fit values. Moreover, the training phase can be terminated by using error criteria. These important properties comprise a structure similar to a tree that is crucial in solving highly complex problems. A set of original inputs are filtered through the layers to the optimal output node when the GMDH network is complete. This is the computational network that must be used in computing predictions (classifications are implied in our application). In GMDH, there are two core concepts: the parametric and the structural identification problems. Figure 4.2 displays the structure of a GMDH model with three model inputs.

Fig. 4.2 Structure of a
GMDH model with three
model inputs

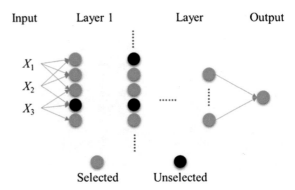

4.3   Tunnel Site and Data Collection

A plan was devised for the effective transfer of water through a tunnel from the
Pahang area to the Selangor area, Malaysia in a way that involved efficient measures
with respect to the problem of water shortage that might occur in the future. The
height of the mountain forming the backbone of Peninsular Malaysia varies between
100 and 1400 m. The plan of the tunnel involved excavation in three sections applying
TBMs and in four sections through common drilling and blasting techniques. A view
of TBM used in this tunnelling project is presented in Fig. 4.3.

Fig. 4.3   TBM used in the study area

Table 4.1 Summary of data used in this study

Variable	Range	Unit	Average	Group
PR	1.11–3.75	m/h	2.37	Output
BTS	4.69–15.1	MPa	9.47	Input
WZ	1–3	–	1.79	Input
RMR	45–95	–	69.3	Input
TF	83–513	kN	279	Input
RPM	4.5–11.9	Rev/min	8.7	Input

For establishing a good database for the prediction TBM PR, there is a need to include at least a component from each important category, i.e., rock mass properties, rock material properties, and machine factors. According to the findings of some researchers (e.g., [61–64]), the extent of mass weathering can influence TBM PR to a considerable degree. Sapigni et al. [65] proposed that the Brazilian tensile strength (BTS) was applicable as an input parameter in forecasting TBM PR. After reviewing the previous relevant studies, we decided to use RMR and weathering zone (WZ) as a component of rock mass properties, BTS as a component of rock material properties, and cutter head thrust force (TF) and revolution per minute (RPM) as a component of machine factors, as input variables to estimate TBM PR. Therefore, we planned to collect block rock samples from the site and then conduct BTS tests in the laboratory. RMR and WZ were observed in tunnel face in different panels (each panel is averagely 10 m). In this site, fresh (with a value of 1 in the modelling), slightly weathered (with a value of 2 in the modelling), and moderately weathered (with a value of 3 in the modelling) were investigated in the study area. The observations and laboratory activities in this study were conducted according to a suggested method by ISRM [66]. A complete version of laboratory tests and field observations in this study can be found in the original study [2].

To predict TBM PR and relying upon a series of laboratory tests and field observations at the tunnel site, a database of 202 datasets was prepared and established in this study. More information such as range, unit, and group regarding the data used in this study are tabulated in Table 4.1. In the following, the mentioned input and output variables will be used in the modelling of AI predictive models.

4.4 GMDH Model Development

This section presents the detailed modelling procedures of GMDH in estimating TBM PR. In this study, the authors decided to use an 80–20% combination for training and testing purposes based on investigations conducted previously [40, 67]. Some key parameters such as the number of GMDH layers, the number of neurons, and the selection pressure must be taken into account in order to design a GMDH model that can predict TBM PR based on five input parameters. A parametric study should be conducted to examine the effects of selection pressure on the system. According to the results obtained, the optimal selection pressure value was 70%. Therefore, 70% as the selection pressure was used in the remaining GMDH modelling. It seems that the number of GMDH layers should be determined and designed through the use of another parametric investigation. To do so, based on the suggestions provided in the literature by several researchers (e.g., [19, 42, 68]), possible numbers of layers were fixed as 2, 5, 8, 12, 14, and 16. Then, six GMDH models were built with the purpose of calculating TBM PR. Table 4.2 shows the results based on the correlation coefficient (R) and mean square error (MSE) of training and testing datasets. It should be noted that this parametric study was performed with six neurons and a 70% selection pressure was set. R and MSE equations are presented in the following:

$$MSE = \frac{1}{M} \sum_{i=1}^{M} (y_{i(Model)} - y_{i(Actual)})^2 \tag{4.1}$$

$$R = \left(\frac{\sum_{i=1}^{M} \left(y_{i(Actual)} - \overline{y}_{(Actual)}\right)\left(y_{i(Model)} - \overline{y}_{(Model)}\right)}{\sum_{i=1}^{M} \left(y_{i(Actual)} - \overline{y}_{(Actual)}\right)^2 \times \sum_{i=1}^{M} \left(y_{i(Model)} - \overline{y}_{(Model)}\right)^2} \right) \tag{4.2}$$

where $y_{i(Model)}$ stands for the predicted PR for each observation ($i = 1, 2, \ldots, M$), $y_{i(Actual)}$ signifies the measured PR values, M represents the number of observations, and $\overline{y}$ is the average of $y_{i(Model)}$ and $y_{i(Actual)}$. A ranking method proposed by Zorlu et al. [69] was used in Table 4.2 and based on the results obtained, model number 4 (i.e., rank value of 22) with 12 layers was selected as the best GMDH model. R and MSE values of 0.958 and 0.0043 and 0.969 and 0.0038 were achieved for train and test of the selected GMDH model.

It is essential to specify the number of neurons using a different parametric investigation when it comes to the last step of GMDH modelling. To do so, a review of the previous studies was conducted (e.g., [42, 68]); and after that, the values of 2, 4, 6, 8, 10, 12, 14, 16, 18, and 20 were selected and used. The results of this parametric study obtained through GMDH models based on R and MSE for the training and testing phases are shown in Table 4.3. As the table shows, the GMDH model that is composed of 14 neurons with the ranking of 34 has the best performance among all the models. The values of 0.961 and 0.959 for R and 0.0041 and 0.0045 for MSE of train and test phases, respectively, were obtained for the best GMDH model. The results confirmed a high ability of this technique in prediction purpose (i.e., TBM PR). This model and its prediction capacity will be discussed later.

Table 4.2 Results of R and MSE for different number of GMDH layers

GMDH model number.	Number of layer	R		MSE		Rank for R		Rank for MSE		Rank summation
		Train	Test	Train	Test	Train	Test	Train	Test	
1	2	0.96	0.959	0.0042	0.0045	6	4	6	4	20
2	5	0.953	0.941	0.0049	0.0063	3	2	2	2	9
3	8	0.958	0.925	0.0045	0.0071	5	1	4	1	11
4	12	0.958	0.969	0.0043	0.0038	5	6	5	6	22
5	14	0.955	0.946	0.0048	0.0055	4	3	3	3	13
6	16	0.958	0.965	0.0043	0.004	5	5	5	5	20

Table 4.3 Results of R and MSE for different number of GMDH neurons

GMDH model number	Number of neuron	R		MSE		Rank for R		Rank for MSE		Rank summation
		Train	Test	Train	Test	Train	Test	Train	Test	
1	2	0.939	0.959	0.006	0.0053	3	9	3	8	23
2	4	0.961	0.957	0.0041	0.0053	8	8	8	8	32
3	6	0.963	0.944	0.004	0.0056	10	5	9	7	31
4	8	0.957	0.935	0.0045	0.0068	6	4	6	4	20
5	10	0.957	0.961	0.0045	0.0039	6	10	6	10	32
6	12	0.96	0.922	0.0044	0.0063	7	2	7	5	21
7	14	0.961	0.959	0.0041	0.0045	8	9	8	9	34
8	16	0.947	0.923	0.0052	0.0097	4	3	4	2	13
9	18	0.954	0.946	0.0047	0.0061	5	6	5	6	22
10	20	0.962	0.951	0.0038	0.007	9	7	10	3	29

4.5 Model Assessment and Discussion

A common action after the development of a predictive model is the assessment of the models according to their performance. A quantitative evaluation of the developed models was conducted using a typical action through the implementation of many performance indices such as R, MSE, error mean, and error StD. These indices have been applied with the aim of evaluating several intelligence relevant studies [29, 70–76]. The error mean and error StD equations are given in the following:

$$E - Mean = \frac{\sum_{i=1}^{M} (y_{i(Actual)} - y_{i(Model)}}{M} \tag{4.3}$$

$$E - StD = \sqrt{\frac{\sum_{i=1}^{M} \left(E_{i(Model)} - \bar{E}_{Model} \right)}{M - 1}} \tag{4.4}$$

in which $y_{i(Model)}$ represents the predicted PR value for each observation ($i = 1, 2, \ldots, M$), $y_{i(Actual)}$ is measured PR value, M is the number of observations, $E_{i(Model)}$ denotes the error value between the measured PR value and predicted one for each data point and $\bar{E}_{Model}$ indicates the mean value of $E_{i(Model)}$.

Figure 4.4 and 4.5 show the results of performance indices obtained for the developed GMDH predictive technique for training and testing phases, respectively. According to these figures, the proposed GMDH model has carried out its

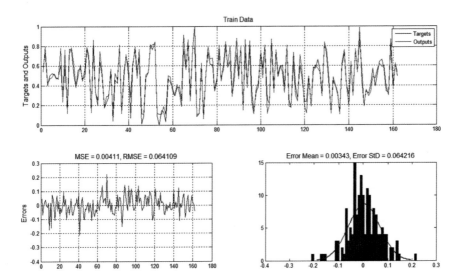

Fig. 4.4 The performance indices result for the train phase of GMDH model

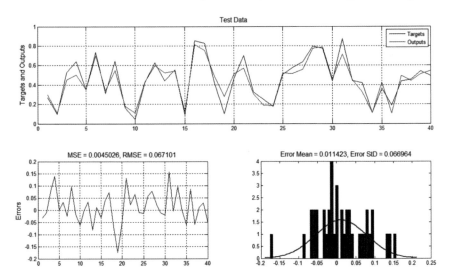

Fig. 4.5 The performance indices result for the test phase of GMDH model

defined predictive tasks acceptably. To have a fair comparison, the authors decided to construct an ANN model for the PR prediction. In this regard, the number of hidden layer and hidden node in the ANN model should be designed. Caudill [77] believes that for mapping a continuous problem, a network with I inputs (i.e., five in this study) should consist of $2I + 1$ hidden neurons (nodes) as the maximum limit. In addition, many studies reported only one hidden layer for an applicable and powerful ANN model [78–81]. Eventually, after considering a range of 1–11 for hidden nodes, the best ANN model was selected with 9 nodes in the hidden layer. In the ANN model, the values of 5, 9, and 1 were obtained for the number of nodes in input, hidden, and output layers, respectively. Results of the ANN model developed in this study for estimating TBM PR are displayed in Figs. 4.6 and 4.7, respectively, for train and test phases. As it can be seen, the ANN is also a powerful technique to do PR prediction task.

The obtained results in Figs. 4.4, 4.5, 4.6 and 4.7 confirm that the GMDH model's precision rate is perceptibly higher than the developed ANN model. Especially, taking the results of testing datasets into account, it was found that the proposed GMDH model can provide higher performance prediction in forecasting TBM PR. For instance, E-StD, MSE, and RMSE values of 0.067, 0.0045, and 0.067 and 0.092, 0.0086 and 0.093 were obtained for the testing phase of GMDH and ANN, respectively, which confirm a higher accuracy level for the GMDH model. It was found that the GMDH predictive model is selected as a powerful method in the area of TBM and tunnelling projects.

It is important to mention that the GMDH model developed in this study includes five input parameters only. This shows that we developed a simpler and easier to conduct predictive model compared to some of the models developed in the previous studies since many of them used more than six or seven input parameters [40, 46, 47,

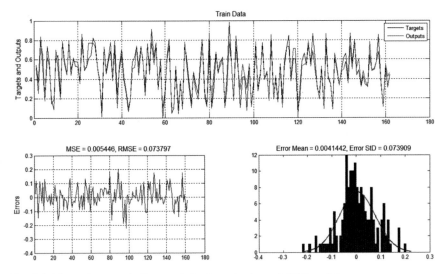

Fig. 4.6 The performance indices result for the train phase of ANN model

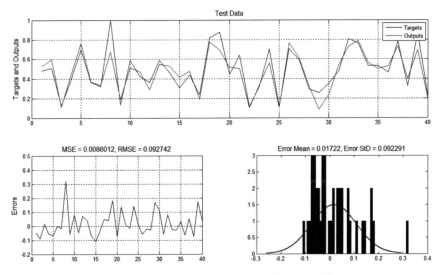

Fig. 4.7 The performance indices result for the test phase of ANN model

82]. Armaghani et al. [83] stated that proposing a simpler model in terms of model inputs is of great importance and possibility although the model performance will be reduced by reducing the model inputs. Therefore, with the obtained accuracy level, it can be concluded that the GMDH model developed in this study is considered as a powerful and applicable technique for both parts of model accuracy and model simplicity.

4.6 Conclusions

The aim of this study was to establish a database for TBM PR prediction on the basis of five input parameters, i.e., RPM, BTS, RMR, TF, and WZ. Compared to similar past studies, we selected and used a lower number of inputs. Then, based on them, several GMDH models, which are actually an updated type of ANN, were developed. The developments of these models were conducted according to their most important factors. For each GMDH effective factor, a separate parametric investigation was carried out. After developing GMDH model, a comparison was made by performing and proposing an ANN model in predicting TBM PR. Then, both models were assessed according to several well-known performance indices. The results of these AI techniques showed that they are capable of realizing a high level of accuracy for the prediction of TBM PR. The results showed that through the development of a GMDH technique, we can get closer PR values to the ones measured in the field compared to PR values predicted by the ANN technique. According to the results of testing data samples, MSE values of 0.0086 and 0.0045 and R values of 0.911 and 0.959 were obtained for the ANN and GMDH models, respectively, which confirm that the GMDH model can predict TBM PR values more accurately than the other techniques employed. The results showed that the GMDH can be proposed and applied as a powerful and practical intelligence model in the area of TBM and tunnelling.

References

1. J. Zhou, Y. Qiu, D.J. Armaghani, W. Zhang, C. Li, S. Zhu, R. Tarinejad, Predicting TBM penetration rate in hard rock condition: a comparative study among six XGB-based metaheuristic techniques. Geosci. Front. (2020). https://doi.org/10.1016/j.gsf.2020.09.020
2. D.J. Armaghani, E.T. Mohamad, M.S. Narayanasamy, N. Narita, S. Yagiz, Development of hybrid intelligent models for predicting TBM penetration rate in hard rock condition. Tunn. Undergr. Sp. Technol. **63**, 29–43 (2017). https://doi.org/10.1016/j.tust.2016.12.009
3. S. Yagiz, C. Gokceoglu, E. Sezer, S. Iplikci, Application of two non-linear prediction tools to the estimation of tunnel boring machine performance. Eng. Appl. Artif. Intell. **22**, 808–814 (2009)
4. F.F. Roxborough, H.R. Phillips, Rock excavation by disc cutter. Int. J. Rock Mech. Min. Sci. Geomech. Abstr. 361–366 (Elsevier, 1975)
5. P.C. Graham, Rock exploration for machine manufacturers. Explor. Rock Eng. 173–180 (1976)
6. I.W. Farmer, N.H. Glossop, Mechanics of disc cutter penetration. Tunnels Tunn. **12**, 22–25 (1980)
7. J. Rostami, Development of a force estimation model for rock fragmentation with disc cutters through theoretical modeling and physical measurement of crushed zone pressure (1997)
8. S. Yagiz, Development of rock fracture and brittleness indices to quantify the effects of rock mass features and toughness in the CSM Model basic penetration for hard rock tunneling machines (2002)
9. P. Bruines, Neuro-fuzzy modeling of TBM performance with emphasis on the penetration rate. Mem. Cent. Eng. Geol. Netherlands Delft. **202** (1998)
10. H.Q. Yang, Z. Li, T.Q. Jie, Z.Q. Zhang, Effects of joints on the cutting behavior of disc cutter running on the jointed rock mass. Tunn. Undergr. Sp. Technol. **81**, 112–120 (2018)

11. H. Yang, H. Wang, X. Zhou, Analysis on the damage behavior of mixed ground during TBM cutting process. Tunn. Undergr. Sp. Technol. **57**, 55–65 (2016)
12. Q.-M. Gong, J. Zhao, Development of a rock mass characteristics model for TBM penetration rate prediction. Int. J. Rock Mech. Min. Sci. **46**, 8–18 (2009)
13. S. Yagiz, H. Karahan, Application of various optimization techniques and comparison of their performances for predicting TBM penetration rate in rock mass. Int. J. Rock Mech. Min. Sci. **80**, 308–315 (2015)
14. A.G. Benardos, D.C. Kaliampakos, Modelling TBM performance with artificial neural networks. Tunn. Undergr. Sp. Technol. **19**, 597–605 (2004)
15. J. Zeng, B. Roy, D. Kumar, A.S. Mohammed, D.J. Armaghani, J. Zhou, E.T. Mohamad, Proposing several hybrid PSO-extreme learning machine techniques to predict TBM performance. Eng. Comput. (n.d.). https://doi.org/10.1007/s00366-020-01225-2
16. J. Zhou, B. Yazdani Bejarbaneh, D. Jahed Armaghani, M.M. Tahir, Forecasting of TBM advance rate in hard rock condition based on artificial neural network and genetic programming techniques. Bull. Eng. Geol. Environ. **79**, 2069–2084 (2020). https://doi.org/10.1007/s10064-019-01626-8
17. D. Jahed Armaghani, P.G. Asteris, B. Askarian, M. Hasanipanah, R. Tarinejad, V. Van Huynh, Examining hybrid and single SVM models with different kernels to predict rock brittleness. Sustainability **12**, 2229 (2020)
18. L. Huang, P.G. Asteris, M. Koopialipoor, D.J. Armaghani, M.M. Tahir, Invasive weed optimization technique-based ann to the prediction of rock tensile strength. Appl. Sci. **9**, 5372 (2019)
19. D.J. Armaghani, E. Momeni, P.G. Asteris, Application of group method of data handling technique in assessing deformation of rock mass. Metaheuristic Comput. Appl. **1**, 1–18 (2020)
20. D.J. Armaghani, P.G. Asteris, A comparative study of ANN and ANFIS models for the prediction of cement-based mortar materials compressive strength. Neural Comput. Appl. (2020). https://doi.org/10.1007/s00521-020-05244-4
21. T.E. Asteris, P.G. Douvika, M.G. Karamani, C.A. Skentou, A.D. Chlichlia, K. Cavaleri, L. Daras, T. Armaghani, D.J. Zaoutis, A novel heuristic algorithm for the modeling and risk assessment of the COVID-19 pandemic phenomenon. Comput. Model. Eng. Sci. (2020). https://doi.org/10.32604/cmes.2020.013280
22. H. Harandizadeh, D.J. Armaghani, Prediction of air-overpressure induced by blasting using an ANFIS-PNN model optimized by GA. Appl. Soft Comput. 106904 (2020)
23. E. Momeni, A. Yarivand, M. Bagher Dowlatshahi, D. Jahed Armaghani, An efficient optimal neural network based on gravitational search algorithm in predicting the deformation of geogrid-reinforced soil structures. Transp. Geotech. 100446 (2020). https://doi.org/10.1016/j.trgeo.2020.100446
24. M. Khari, D.J. Armaghani, A. Dehghanbanadaki, Prediction of lateral deflection of small-scale piles using hybrid PSO–ANN Model. Arab. J. Sci. Eng. (2019). https://doi.org/10.1007/s13369-019-04134-9
25. J. Huang, M. Koopialipoor, D.J. Armaghani, A combination of fuzzy Delphi method and hybrid ANN-based systems to forecast ground vibration resulting from blasting. Sci. Rep. **10**, 1–21 (2020)
26. B.R. Murlidhar, D. Kumar, D. Jahed Armaghani, E.T. Mohamad, B. Roy, B.T. Pham, A novel intelligent ELM-BBO technique for predicting distance of mine blasting-induced flyrock. Nat. Resour. Res. (2020). https://doi.org/10.1007/s11053-020-09676-6
27. D. Ramesh Murlidhar, B. Yazdani Bejarbaneh, B. Jahed Armaghani et al., Application of tree-based predictive models to forecast air overpressure induced by mine blasting. Nat. Resour. Res. (2020). https://doi.org/10.1007/s11053-020-09770-9
28. M. Hajihassani, S.S. Abdullah, P.G. Asteris, D.J. Armaghani, A gene expression programming model for predicting tunnel convergence. Appl. Sci. **9**, 4650 (2019)
29. D.J. Armaghani, P.G. Asteris, S.A. Fatemi, M. Hasanipanah, R. Tarinejad, A.S.A. Rashid, V. Van Huynh, On the use of neuro-swarm system to forecast the pile settlement. Appl. Sci. **10**, 1904 (2020)

30. H. Chen, P.G. Asteris, D. Jahed Armaghani, B. Gordan, B.T. Pham, Assessing dynamic conditions of the retaining wall: developing two hybrid intelligent models. Appl. Sci. **9**, 1042 (2019)
31. J. Zhou, P.G. Asteris, D.J. Armaghani, B.T. Pham, Prediction of ground vibration induced by blasting operations through the use of the Bayesian Network and random forest models. Soil Dyn. Earthq. Eng. **139**, 106390 (2020). https://doi.org/10.1016/j.soildyn.2020.106390
32. S. Lu, M. Koopialipoor, P.G. Asteris, M. Bahri, D.J. Armaghani, A novel feature selection approach based on tree models for evaluating the punching shear capacity of steel fiber-reinforced concrete flat slabs. Mater. (Basel) **13**, 3902 (2020)
33. P.G. Asteris, D.J. Armaghani, G.D. Hatzigeorgiou, C.G. Karayannis, K. Pilakoutas, Predicting the shear strength of reinforced concrete beams using artificial neural networks. Comput. Concr. **24**, 469–488 (2019)
34. M. Apostolopoulou, P.G. Asteris, D.J. Armaghani, M.G. Douvika, P.B. Lourenço, L. Cavaleri, A. Bakolas, A. Moropoulou, Mapping and holistic design of natural hydraulic lime mortars. Cem. Concr. Res. **136**, 106167 (2020)
35. P. Sarir, J. Chen, P.G. Asteris, D.J. Armaghani, M.M. Tahir, Developing GEP tree-based, neuroswarm, and whale optimization models for evaluation of bearing capacity of concrete-filled steel tube columns. Eng. Comput. (2019). https://doi.org/10.1007/s00366-019-00808-y
36. M. Koopialipoor, H. Tootoonchi, D. Jahed Armaghani, E. Tonnizam Mohamad, A. Hedayat, Application of deep neural networks in predicting the penetration rate of tunnel boring machines. Bull. Eng. Geol. Environ. (2019). https://doi.org/10.1007/s10064-019-01538-7
37. S. Yagiz, H. Karahan, Prediction of hard rock TBM penetration rate using particle swarm optimization. Int. J. Rock Mech. Min. Sci. **48**, 427–433 (2011)
38. M.G. Simoes, T. Kim, Fuzzy modeling approaches for the prediction of machine utilization in hard rock tunnel boring machines, in *Inductive Applied Conference 2006, 41st IAS Annual Meeting Conference Record 2006 IEEE* (IEEE, 2006), pp. 947–954
39. J. Zhou, Y. Qiu, S. Zhu, D.J. Armaghani, M. Khandelwal, E.T. Mohamad, Estimation of the TBM advance rate under hard rock conditions using XGBoost and Bayesian optimization. Undergr. Sp. (2020). https://doi.org/10.1016/j.undsp.2020.05.008
40. D.J. Armaghani, M. Koopialipoor, A. Marto, S. Yagiz, Application of several optimization techniques for estimating TBM advance rate in granitic rocks. J. Rock Mech. Geotech. Eng. (2019). https://doi.org/10.1016/j.jrmge.2019.01.002
41. J. Zhou, Y. Qiu, S. Zhu, D.J. Armaghani, C. Li, H. Nguyen, S. Yagiz, Optimization of support vector machine through the use of metaheuristic algorithms in forecasting TBM advance rate. Eng. Appl. Artif. Intell. **97**(n.d.), 104015 (2021)
42. M. Koopialipoor, S.S. Nikouei, A. Marto, A. Fahimifar, D.J. Armaghani, E.T. Mohamad, Predicting tunnel boring machine performance through a new model based on the group method of data handling. Bull. Eng. Geol. Environ. **78**, 3799–3813 (2018)
43. M. Koopialipoor, A. Fahimifar, E.N. Ghaleini, M. Momenzadeh, D.J. Armaghani, Development of a new hybrid ANN for solving a geotechnical problem related to tunnel boring machine performance. Eng. Comput. (2019). https://doi.org/10.1007/s00366-019-00701-8
44. H. Xu, J. Zhou, P. G Asteris, D. Jahed Armaghani, M.M. Tahir, Supervised machine learning techniques to the prediction of tunnel boring machine penetration rate. Appl. Sci. **9**, 3715 (2019)
45. D.J. Armaghani, R.S. Faradonbeh, E. Momeni, A. Fahimifar, M.M. Tahir, Performance prediction of tunnel boring machine through developing a gene expression programming equation. Eng. Comput. **34**, 129–141 (2018)
46. M.A. Grima, P.A. Bruines, P.N.W. Verhoef, Modeling tunnel boring machine performance by neuro-fuzzy methods. Tunn. Undergr. Sp. Technol. **15**, 259–269 (2000)
47. S. Mahdevari, K. Shahriar, S. Yagiz, M.A. Shirazi, A support vector regression model for predicting tunnel boring machine penetration rates. Int. J. Rock Mech. Min. Sci. **72**, 214–229 (2014)
48. H. Fattahi, Adaptive neuro fuzzy inference system based on fuzzy c-means clustering algorithm, a technique for estimation of tbm penetration rate. Iran Univ. Sci. Technol. **6**, 159–171 (2016)

49. V.T. Minh, D. Katushin, M. Antonov, R. Veinthal, Regression models and fuzzy logic prediction of tbm penetration rate. Open Eng. **7**, 60–68 (2017)
50. Y. Won, S. Han, D. Seong, Bearing capacity and settlement of tire-reinforced sands. **22**, 439–453 (2004). https://doi.org/10.1016/j.geotexmem.2003.12.002
51. E.T. Mohamad, R.S. Faradonbeh, D.J. Armaghani, M. Monjezi, M.Z.A. Majid, An optimized ANN model based on genetic algorithm for predicting ripping production. Neural Comput. Appl. **28**, 393–406 (2017)
52. Rosenblatt F, The perceptron: a probabilistic model for information storage and organization in the brain. Psychol. Rev. **65**(386) (1958)
53. P.G. Asteris, M. Apostolopoulou, A.D. Skentou, A. Moropoulou, Application of artificial neural networks for the prediction of the compressive strength of cement-based mortars. Comput. Concr. **24**, 329–345 (2019)
54. E.T. Mohamad, D.J. Armaghani, E. Momeni, A.H. Yazdavar, M. Ebrahimi, Rock strength estimation: a PSO-based BP approach. Neural Comput. Appl. **30**, 1635–1646 (2018)
55. H. Nguyen, X.N. Bui, Y. Choi, C.W. Lee, D.J. Armaghani, A novel combination of whale optimization algorithm and support vector machine with different kernel functions for prediction of blasting-induced fly-rock in quarry mines. Nat. Resour. Res. (2020). https://doi.org/10.1007/s11053-020-09710-7
56. I.A. Basheer, M. Hajmeer, Artificial neural networks: fundamentals, computing, design, and application. J. Microbiol. Methods **43**, 3–31 (2000)
57. E.T. Mohamad, S.A. Noorani, D.J. Armaghani, R. Saad, Simulation of blasting induced ground vibration by using artificial neural network. Elect. J. Geotech. Eng. **17**, 2571–2584 (2012)
58. D.J. Armaghani, R.S.N.S. Bin Raja, K. Faizi, A.S.A. Rashid, Developing a hybrid PSO–ANN model for estimating the ultimate bearing capacity of rock-socketed piles. Neural Comput. Appl. **28**, 391–405 (2017)
59. M. Hajihassani, D. Jahed Armaghani, H. Sohaei, E. Tonnizam Mohamad, A. Marto, Prediction of airblast-overpressure induced by blasting using a hybrid artificial neural network and particle swarm optimization. Appl. Acoust. **80**, 57–67 (2014). https://doi.org/10.1016/j.apacoust.2014.01.005
60. A.G. Ivakhnenko, Polynomial theory of complex systems. IEEE Trans. Syst. Man Cybern. **1**, 364–378 (1971)
61. M. Sundaram, The effects of ground conditions on TBM performance in tunnel excavation–A case history (2007)
62. N.M. Sundaram, A.G. Rafek, I. Komoo, The influence of rock mass properties in the assessment of TBM performance, in *Proceeding of 8th IAEG Congress* (Vancouver, Br. Columbia, Canada, 1998), pp. 3553–3559
63. W. Shijing, Q. Bo, G. Zhibo, The time and cost prediction of tunnel boring machine in tunnelling. Wuhan Univ. J. Nat. Sci. **11**, 385–388 (2006)
64. S. Yagiz, Utilizing rock mass properties for predicting TBM performance in hard rock condition. Tunn. Undergr. Sp. Technol. **23**, 326–339 (2008)
65. M. Sapigni, M. Berti, E. Bethaz, A. Busillo, G. Cardone, TBM performance estimation using rock mass classifications. Int. J. Rock Mech. Min. Sci. **39**, 771–788 (2002)
66. R. Ulusay, J.A. Hudson, ISRM The complete ISRM suggested methods for rock characterization, testing and monitoring: 1974–2006. Comm. Test. Methods. Int. Soc. Rock Mech. Compil. **628**(n.d.) (Arranged by ISRM Turkish Natl. Group, Ankara, Turkey 2007)
67. E. Momeni, D.J. Armaghani, S.A. Fatemi, R. Nazir, Prediction of bearing capacity of thin-walled foundation: a simulation approach. Eng. Comput. **34**, 319–327 (2018)
68. D. Li, M.R. Moghaddam, M. Monjezi, D. Jahed Armaghani, A. Mehrdanesh, Development of a group method of data handling technique to forecast iron ore price. Appl. Sci. **10**, 2364 (2020)
69. K. Zorlu, C. Gokceoglu, F. Ocakoglu, H.A. Nefeslioglu, S. Acikalin, Prediction of uniaxial compressive strength of sandstones using petrography-based models. Eng. Geol. **96**, 141–158 (2008)

70. D.J. Armaghani, F. Mirzaei, M. Shariati, N.T. Trung, M. Shariati, D. Trnavac, Hybrid ANN-based techniques in predicting cohesion of sandy-soil combined with fiber. Geomech. Eng. **20**, 191–205 (2020)

71. H. Harandizadeh, D.J. Armaghani, E.T. Mohamad, Development of fuzzy-GMDH model optimized by GSA to predict rock tensile strength based on experimental datasets. Neural Comput. Appl. **32**, 14047–14067 (2020). https://doi.org/10.1007/s00521-020-04803-z

72. D.J. Armaghani, M. Koopialipoor, M. Bahri, M. Hasanipanah, M.M. Tahir, A SVR-GWO technique to minimize flyrock distance resulting from blasting. Bull. Eng. Geol. Environ. (2020). https://doi.org/10.1007/s10064-020-01834-7

73. D. Tang, B. Gordan, M. Koopialipoor, D. Jahed Armaghani, R. Tarinejad, B. Thai Pham, V. Van Huynh, Seepage analysis in short embankments using developing a metaheuristic method based on governing equations. Appl. Sci. **10**, 1761 (2020)

74. J. Ye, J. Dalle, R. Nezami, M. Hasanipanah, D.J. Armaghani, Stochastic fractal search-tuned ANFIS model to predict blast-induced air overpressure. Eng. Comput. (2020). https://doi.org/10.1007/s00366-020-01085-w

75. Z. Yu, X. Shi, J. Zhou, Y. Gou, X. Huo, J. Zhang, D.J. Armaghani, A new multikernel relevance vector machine based on the HPSOGWO algorithm for predicting and controlling blast-induced ground vibration. Eng. Comput. (2020). https://doi.org/10.1007/s00366-020-01136-2

76. W. Yong, J. Zhou, D.J. Armaghani, M.M. Tahir, R. Tarinejad, B.T. Pham, V. Van Huynh, A new hybrid simulated annealing-based genetic programming technique to predict the ultimate bearing capacity of piles. Eng. Comput. (2020). https://doi.org/10.1007/s00366-019-00932-9

77. M. Caudill, Neural networks primer. Part III AI Expert. **3**, 53–59 (1988)

78. S.V. Alavi Nezhad Khalil Abad, M. Yilmaz, D. Jahed Armaghani, A. Tugrul, Prediction of the durability of limestone aggregates using computational techniques. Neural Comput. Appl. (2016). https://doi.org/10.1007/s00521-016-2456-8

79. E.T. Mohamad, D.J. Armaghani, M. Hajihassani, K. Faizi, A. Marto, A simulation approach to predict blasting-induced flyrock and size of thrown rocks. Electron. J. Geotech. Eng. **18**(B), 365–374 (2013)

80. D.J. Armaghani, E.T. Mohamad, M. Hajihassani, S. Yagiz, H. Motaghedi, Application of several non-linear prediction tools for estimating uniaxial compressive strength of granitic rocks and comparison of their performances. Eng. Comput. **32**, 189–206 (2016)

81. B.Y. Bejarbaneh, E.Y. Bejarbaneh, A. Fahimifar, D.J. Armaghani, M.Z.A. Majid, Intelligent modelling of sandstone deformation behaviour using fuzzy logic and neural network systems. Bull. Eng. Geol. Environ. **77**, 345–361 (2018)

82. M. Eftekhari, A. Baghbanan, M. Bayati, Predicting penetration rate of a tunnel boring machine using artificial neural network, in *ISRM International Symposium Asian Rock Mechanics Symposium* (International Society for Rock Mechanics, 2010)

83. D.J. Armaghani, E.T. Mohamad, E. Momeni, M.S. Narayanasamy, An adaptive neuro-fuzzy inference system for predicting unconfined compressive strength and Young's modulus: a study on Main Range granite. Bull. Eng. Geol. Environ. **74**, 1301–1319 (2015)

Printed in the United States
by Baker & Taylor Publisher Services